Reporting Results

A PRACTICAL GUIDE FOR ENGINEERS AND SCIENTISTS

David C. Van Aken
Missouri University of Science
and Technology

William F. Hosford
University of Michigan

CAMBRIDGE
UNIVERSITY PRESS

CAMBRIDGE UNIVERSITY PRESS
Cambridge, New York, Melbourne, Madrid, Cape Town, Singapore,
São Paulo, Delhi

Cambridge University Press
32 Avenue of the Americas, New York, NY 10013–2473, USA

www.cambridge.org
Information on this title: www.cambridge.org/9780521723480

First published 2008

Printed in the United States of America

A catalog record for this publication is available from the British Library.

Library of Congress Cataloging in Publication Data

Van Aken, David C.
Reporting results : a practical guide for engineers and scientists /
David C. Van Aken, William F. Hosford.
    p.   cm.
Includes bibliographical references and index.
ISBN 978-0-521-89980-2 (hardback) – ISBN 978-0-521-72348-0 (pbk.)
1. Technical writing.   2. Communication of technical
information.   I. Hosford, William F.   II. Title.
T11.V358   2008
808′.0666–dc22   2007051436

ISBN   978-0-521-89980-2 hardback
ISBN   978-0-521-72348-0 paperback

Contents

Preface

This brief guide was written for science and engineering students and professionals to help them communicate technical information clearly, accurately, and effectively. The focus is on the most common communication forms and the most common issues that arise in classroom and professional practice.

Freshman chemistry or physics will be the introduction to technical report writing for many college students. The format for writing these laboratory reports is most often specified by the instructor. This guide will be useful in developing a good technical writing style and for preparing tables and figures for those reports. Upper-level courses often use the same formatting as is required for submission to technical journals or for technical report writing, which is the focus of this book. Graduate students and professionals encounter many of the same problems in technical communication. Good communication skills are required in all forms of technical writing and presentation. This book is designed to help the reader develop effective communication skills and to be a reference on stylistic and grammar issues. Unlike most texts on writing style, this book also treats oral presentations, graphing, and analysis of data.

The authors' intention is to give the reader the basics of technical communication in the first chapter and then to treat in detail the various forms of technical communication. The structure of the book is as follows:

Chapter 1 provides a general discussion of technical communication.

Chapter 2 covers writing technical reports and archival papers.

Chapter 3 discusses writing letter reports, which are common in industry.

Chapter 4 gives general guidelines for oral presentations.

Chapter 5 treats the effective use of tables and figures, with an emphasis on the science of graphing.

Chapter 6 covers some basic concepts in the statistical analysis of data.

Chapter 7 offers suggestions for writing resumés.

The appendices treat common errors in writing, including punctuation and commonly confused words; general information, including the international system of numerical prefixes and units and the Greek alphabet; and uses of straight lines to represent some mathematical functions.

This guide is intended for all science and engineering majors. The careful reader may notice that many of the examples are taken from the authors' experiences in materials science and engineering.

Clear communication is a challenge that often does not appeal to engineers and scientists. However, the responsibility of ethical scientists and engineers is to ensure that humanity benefits from their knowledge. If one is unable to communicate one's ideas effectively, then for all practical purposes the work is lost. Academic grades and future careers are dependent on good communication skills. Becoming a good writer is a lifelong journey, and the authors hope that this book provides a quick reference for readers in both their academic and their professional careers.

REPORTING RESULTS

1 Elements of Technical Writing

The ability to communicate clearly is the most important skill engineers and scientists can have. Their best work will be lost if it is not communicated effectively. In this chapter, elements of the technical style of writing are examined. Technical writing differs in presentation and tone from other styles of writing; these differences are described first. The most important elements of the technical writing style to be discussed are conciseness and unambiguity. The chapter ends with a discussion of proofreading and some helpful hints in developing technical writing skills.

Presentation and Tone

Technical communication differs from fiction in many ways. In mystery novels the reader is kept in suspense because the writer has hidden important clues that are explained at the end of the story to produce a surprise. In contrast, the readers of technical writing are given the important conclusions at the beginning, followed by evidence supporting those conclusions. The following example illustrates the difference. The simple question *Do we have any mail today?* can be answered by a man sitting on his porch in two ways.

He could say: "I got up out of my chair and sauntered out to the mailbox. I looked up before opening the box and saw the mailman going down the street past our house. When I opened the mailbox there was nothing in it, so I don't think we'll have any mail today."

Or he could answer: "No, we won't have mail today. The mailbox is empty and the mailman has passed our house."

Note that in the first reply, the reader must wait until the end of the story to find the answer. This is typical of fiction writing. In the second reply the answer is given up front and then justified. The tone of the second reply is kept factual. This is what technical writing should do.

Number, Voice, and Tense

Most technical communication is done in the third person. Pronouns like *you*, *I*, and *we* are to be avoided. Only Nobel laureates may write in the first person without seeming to be pompous.

Readers probably studied voice in an English class. As a reminder, examples of the different types of voice are:

Active voice: The ice melted at 0°C.

Passive voice: The ice was melted by convection heating.

Imperative voice: Place the ice in a convection oven until the ice melts.

The imperative voice is seldom used in technical communication except when giving instructions about how to do something. It tends to sound like the author is ordering the reader to do something. There is a strong temptation to overuse the passive voice in technical writing to avoid using *I* and *we*; however, it is good to use the active voice wherever possible.

Past and perfect tenses are used in technical writing, because they are used to report something that has happened. The difference in tenses is illustrated by the following:

Past tense: A break in the circuit interrupted the current.

Perfect tense: A break in the circuit has interrupted the current.

It is usually best to pick a tense and be consistent with it in your writing. Frequent shifting of tenses can leave the reader confused. Occasionally, the past perfect tense can be used to describe a prior event. The previous example written in the past perfect tense is "A break in the circuit had interrupted the current." An exception when it is okay to use the present tense is when stating an enduring truth like "Current passing through a resistor causes it to heat up."

Conciseness

A hallmark of good technical papers and reports is that they are as concise as is consistent with being complete and unambiguous. Most readers are busy people, and the writer should avoid wordiness and redundancy. In writing a technical report, one can often assume that the audience is familiar with the scientific and engineering terminology.

Consider the following excerpt from the middle of a doctoral thesis proposal.

> A schematic illustration of the spot friction welding process is shown in Figure 1. The process is applied to join the two metal sheets as shown. A rotating tool with a probe pin is first plunged into the upper sheet. When the rotating tool contacts the upper sheet, a downward force is applied. A backing plate beneath the lower sheet is used to support the downward force of the tool. The downward force and the rotational speed of the tool are maintained for an appropriate

time to generate frictional heat. Then, heated and softened material adjacent to the tool deforms plastically, and a solid-state bond is made between the surfaces of the upper and lower sheets. Finally, the tool is drawn out of the sheets as shown in Figure 1.

This could be rewritten in a much more concise form without any loss of meaning as:

Figure 1 is a schematic illustration of spot friction welding of two sheets. A rotating tool with a probe pin is plunged into the upper sheet. A backing tool beneath the lower sheet supports the downward force of the tool. The force and rotational speed are maintained long enough to generate heat. The heated material adjacent to the tool deforms plastically and forms a solid-state bond.

Note that the number of words is reduced from 130 to 66. The word *process* is unnecessary and overused. *Welding process* means welding; likewise *machining process* means machining and *rolling process* means rolling. The fact that the force is applied to the rotating tool is obvious, that the bond is between the upper and lower sheets is also obvious, and saying that the tool is removed is not needed.

Another example is taken from a draft of a doctoral thesis.

For decades, study of fracture has been one of many important topics that have attracted enough interest, due to its implication in a wide range of practical and real-life problems. Automotive safety, for one example, has been a focus and major challenge facing the industry. Structural integrity and failure is one of the key areas that are closely connected to automotive safety. For instance, in a rear car-to-car crash event, the main concern on the recipient car is fuel system integrity, or in other words, the condition of the fuel tank and fuel pipes. If the rear structure of the recipient car fails to protect the fuel tank or the fuel pipe from being crushed or punctured, any subsequent crack in the fuel tank or fuel pipe will lead to fuel leakage, which poses immediate danger of fire burst. A full-vehicle finite element (FE) model with well over one million elements and tens of millions degrees of freedom is currently developed to help vehicle design within the automotive industry. These large-scale models can now be solved on a group of high-speed workstations or small computers by the process of multi-process parallel computing. However, given such a large-scale model, the failure or failed component may well be within a small or localized area; as aforementioned, a crack in the fuel tank or fuel pipe. It is still lacking in an FE model to simulate reasonably how a crack initiates and grows under impact loading.

This paragraph could be shortened to:

Structural integrity is closely related to automotive safety. For example, in an end-to-end crash, the fuel system integrity is crucial. Full-vehicle finite element models (FEM) with more than a million elements and ten million degrees of freedom are currently being developed to aid vehicle design. However, these models do not reasonably simulate crack initiation and growth under impact loading.

Unambiguity

Technical writing should be unambiguous so the audience knows exactly what the writer intends.

Consider the following paragraph from a student report:

> The cast is removed from the oven and molten metal was poured into the mold until the sprue filled. The mold is cooled until the metal is no longer red-hot. It is then placed into a water bath to remove the investment from the casting.

Here there is confusion between the words *cast*, *mold*, and *investment*. It is not clear what is meant by the word *cast*. It is unnecessary to refer to filling of the *sprue*, and the word *bath* adds nothing. A shorter and more precise version might be:

> Molten metal is poured into a preheated investment mold. When the casting is no longer hot, it is plunged into water, facilitating removal of the investment.

Consider the sentence *A gray iron casting consists of a steel matrix with a flake-graphite phase, which can come out of solution and lower the density of the final casting.* The wording *Gray iron consists of a steel matrix with a flake-graphite phase, which lowers the density of the final casting* would be clearer and more concise.

Another form of ambiguity is when a writer refers, for example, to a *copper aluminum alloy*. This could be interpreted either as a *copper-base aluminum alloy* or a *copper-containing aluminum-base alloy*.

Reflexive pronouns should be used carefully in technical writing. The antecedent to words like *that, which, he, she,* and *it* must be unambiguous. In the sentence *The addition of magnesium to iron above that of its boiling point converted it to a ductile state,* the antecedents of *its* and *it* are unclear. The reader will not be able to tell whether magnesium is added above the boiling point of iron or of magnesium and whether iron or magnesium is made ductile.

Another common source of ambiguity is to refer the reader to a table of data or a figure without explaining what the reader is to learn from looking at it. Good figure captions can eliminate some of these problems, but they are not a substitute for good writing.

In all cases, the writer must make clear what work was done by the author and what was learned from the literature. Citations like *It is known that . . .* , *They believe that . . .* , and *Most engineers agree that . . .* should be avoided unless specific references are given.

Use of Acronyms

Occasional use of acronyms can reduce the number of required words. Each acronym should be introduced by writing the full phrase out. For example, *scanning electron microscopy (SEM)*. The number of acronyms used in a paper should be kept to a minimum. There is no need to introduce an acronym if the term is only used two or three times. Each acronym requires the reader to learn a new bit of jargon, which can make reading more cumbersome.

An example of the excessive use of acronyms is *Compared to other SPF processes, the ABRC process offers the possibility for scaling up the production of UFG Mg sheets. The high pressure of the ECAE process is also avoided.* This could be rewritten as *Alternate biaxial reverse compression can be used to produce large quantities of ultra-fine-grain Mg for superplastic forming while avoiding the high pressures required in equal channel angle extrusion.*

Proofreading

All writing should be proofread, with the specific aims of checking for grammar, spelling, and word errors; eliminating repetition of words and ideas; checking the flow of thoughts; and seeing if sentence length is varied. Reading a manuscript aloud or rehearsing an oral presentation to oneself will reveal clumsy wording. For example, the sentence *The radiograph was used to show where defects were in the specimens, such as voids and porosity* would read better as *The radiograph was used to show where defects, such as voids and porosity, were in the specimens.*

There are different rules to follow for using adjectives and adverbs. Adjectives modify only nouns. Adverbs may modify either verbs or adjectives. One common misuse is found on the traffic sign that says *Drive Slow* instead of *Drive Slowly*. *Slow* is an adjective whereas *slowly* is an adverb, so *Drive Slow* is grammatically incorrect. In technical writing, both *slow heating* and *heating slowly* are correct. In the first case, *heating* is used as a noun, and in the second *heating* is used as a verb.

Agreement between subject and predicate can be checked by leaving out all words between them. For example, *A group of circuits with resistors, capacitors, and other circuit elements are shown...* may sound correct but when read as *A group are shown*, the need to change *are* to *is* becomes apparent.

Contractions of words, such as it's for it is and can't for cannot, are acceptable in common speech and fiction writing, but they should not be used in technical writing.

The word *then* is often overused in technical writing when describing procedures. For example, the sentences *The model was constructed and then painted with a waterproof paint. Then it was placed in a tow tank for testing* could be replaced by *After construction, the model was painted with a waterproof paint and placed in a tow tank for testing.* Other examples of excess verbiage are given in Appendix I.

Poor choice of words, awkward phrasing, and misused words will distract the reader and lower your credibility as an author. A common example of bad phrasing is *very unique.* If something is *unique* it means that there is nothing else like it. Something is either unique or it is not; it cannot be *very unique.* Also, *many* or *much* is better than *lots of* or *loads of.* Examples of common errors with words are discussed in Appendix I. Punctuation is explained in Appendix II, and commonly confused words are listed in Appendix III.

Varying the length of sentences makes reading less monotonous than if all of the sentences are either short or long. Sentences that are too lengthy can often be broken into two or three separate sentences. For example, the sentence *In rear-end crashes, the leading car may suffer fuel system damage leading to leakage, which may lead to fire, injuring the passengers or even causing death* could be rewritten as *Rear-end crashes may damage the fuel system of the lead car. Fire may result from gas leakage, and this can result in injury or death.* Repeating the same word too many times when a synonym could be used is also monotonous.

Chapter Summary

Technical writing should convey the most important findings first. It should be in the third person, factual, and concise. Writing should be unambiguous. It is important that both the author and others proofread the work; it is always good to have an extra pair of eyes checking for errors. Finally, criticism of one's work should be accepted as an opportunity for improvement.

2 Technical Papers

Technical papers are a principal means of communicating within the scientific community. They are generally archival in nature and follow prescribed formats dependent on the journal or publisher. Laboratory instructors may require a format similar to a technical journal, and students will find this chapter useful in preparing their technical reports. Corporations and government agencies may have different requirements; these are not addressed here. This chapter describes various formats and describes how the general subsections – abstract, background, experimental procedures, results, discussion, summary, acknowledgments, and references – should be written.

Format

There are various formats that can be used for technical papers. The format should use headings and subheadings that divide the text into convenient portions. Formats are designed for optimum communication to the reader and can provide easily recognizable locations in the text to which the reader can return after interruption. Also, important results can be associated with specific headings, helping the reader find information of interest. Although no set format is best for all technical reports, all formats require concise but complete documentation.

A simple format for a technical paper or report contains the following: title, abstract, introduction, results, discussion, conclusions, acknowledgments, references, and appendices. The title page contains the title of the paper and the authors' names and affiliations. Any figures and tables should be incorporated into the body of the text as soon after they are referred to as is convenient, or they can be collected at the end of the report. Figures and tables should be limited to those necessary to justify the conclusions of the report. References should appear at the end of the paper; however, if five or fewer are used, they may be incorporated as footnotes. There is no need for a table of contents unless the report covers more than 50 pages.

Other formats may be used where appropriate. Table 2.1 shows three basic formats used for journal papers.

Table 2.1. *Common journal formats for papers*

Style 1	Style 2	Style 3
Title	Title	Title
Abstract	Summary	Abstract
Introduction	Background	Introduction
Experimental Procedures	Theoretical Model	Experimental Procedures
Results	Results	Results and Discussion
Discussion	Discussion	Summary
Summary	Conclusions	Acknowledgments
Acknowledgments	Acknowledgments	References
References	References	
Appendices	Appendices	

Sometimes the nature of an article calls for a different format. A few formats from articles in different journals are listed here:

From *International Journal of Mechanical Sciences*

Title: "Influence of Strain-Path Changes on Forming Limit Diagrams of Al 6111 T4"

Format: Abstract, notation, introduction, experimental procedure, results, forming limits in plane-strain after various prestrain paths, industrial observations, conclusions, references

From *Journal of Applied Physics*

Title: "Operation of Bistable Phase-Locked Single-Electron Tunneling Logic Elements"

Format: Abstract, introduction, principle of operation, model, bistability of an isolated gate, return map, locking of a single-gate to a sinusoidal input signal, interaction between coupled gates, signal transfer in separated clock stages, circuit implementation, conclusions, acknowledgments, references

From *Journal of Aerospace Science*

Title: "Large-Eddy Simulation"

Format: Abstract, introduction, formulation, sub-grid scale models, numerical methods, achievements, challenges, conclusion, acknowledgments, references

From *AIChE Journal*

Title: "Penetration of Shear Flow into an Array of Rods Aligned with the Flow"

Format: Abstract, introduction, prior work and objective, validation of technique, results for shear-driven flow, conclusions, acknowledgments, appendices, references

From *Journal of the Astronautical Sciences*

Title: "Tracking Rigid Body Motion Using Thrusters and Momentum Wheels"

Format: Abstract, introduction, system models, dynamics, kinematics, tracking controllers, numerical examples, conclusions, appendices, acknowledgments, references

Note that all of these formats start with an abstract (or summary), followed by an introduction to the subject material, and end with conclusions, acknowledgments, and references. Appendices, if any, are at the end.

Title

Titles should be short and not too general or specific. For example, the title "Analysis and Comparison of the Transportation Systems of Several Major Cities" could be shortened to "Analysis of Urban Transportation Systems." Note that *Analysis and Comparison* can be replaced by *Analysis*, and *Several Major Cities* can be replaced by *Urban*. However, shortening the title to "Analysis of Transportation Systems" would incorrectly imply inclusion of air, rail, and ship transportation.

Abstract

The abstract is a concise summary of the significant items in a report. Typically, an abstract contains between 200 and 400 words. It should include what has been studied, significant results, and conclusions. Simply stating that transportation systems were analyzed is insufficient. The results of the analysis should also be stated. For example, *In analyzing the transportation systems, it was found that subways are the most efficient means of transporting large numbers of people.* The abstract should also report significant findings. For example, rather than state *Radiation pressure was measured using a torsion balance technique,* write *Using a torsion balance technique, radiation pressure was measured to be* 7.01×10^{-6} nt/m^2 *versus a predicted value of* 7.05×10^{-6} nt/m^2. In combination with the title, the abstract should indicate the content of the report. Abstracts of technical papers are often published separately. Therefore the abstract must be able to stand alone without reference to figures, tables, or anything else in the body of the paper.

Introduction

This section introduces the reader to the topic of the report. The introduction contains the objectives of the paper and important background information. Pertinent literature may be surveyed. The introduction usually ends with a very specific statement of purpose.

Experimental Procedures

This section describes the experimental methods, including the materials used and specific procedures followed. The description of the experimental procedure should allow the reader to evaluate and reproduce the experiment. Thus, for both the credibility of the results and future developments based on the results, this section is very important. In the interest of conciseness, details of standard procedures and equipment should be omitted. Statistical methods used to analyze the precision or experimental errors should be explained. It should be clear to the reader what was done by the author(s) and what was done by others.

Results

The outcomes of the experiments are reported in this section. The results should be arranged in a logical sequence appropriate to the experiment and should include pertinent figures and tabulated data. The order of presentation need not always correspond to the chronological order of the tasks.

Discussion

Next, the results are analyzed and interpreted. The results should be discussed in context with the prior work reported in the introduction. Sometimes results and discussion are combined into a single section, but this practice can lead to ambiguity. It should be made clear what is new and what is from previous work.

Summary or Conclusions

A summary or statement of conclusions should always be included at the end of the report to provide closure. Often a busy reader will turn to this section before deciding whether to read the paper.

Acknowledgments

Where appropriate, ideas, sources of financial aid, and help from others should be acknowledged.

References

Statements of fact and citations of prior work should be referenced in the text so the reader can access the original work. Most research and development projects rely on the results of other projects reported in journals and reports. A reference number should appear in the text so that it is clear to the reader what is being referenced. In citing prior work, use a reference number and the name of the author(s). If there are more than two authors, cite the first author and "et al." Reference numbers in the text should be in sequential order and enclosed in brackets or superscripted. Some journals may require different formats. However, one should be consistent and use only one style in each report. Each reference at the back of the report should contain all of the pertinent information that will allow the reader to find the cited work. This includes the author(s), the title of the work, the source (journal or book title), the date, and the page numbers. The reference section is not a bibliography. It should only contain references directly cited in the report text.

A suggested form is: [Reference number] Name of Author(s), "Title of Article," *Name of Publication,* Vol. No., Publisher Name and Location (date of publication), page numbers. Some specific examples are listed here:

Journal article:	[1] J. H. Smith and E. R. Vance, "Decomposition of Gamma-Phase Manganese Copper Alloy," *J. Appl. Phys.* 40 (1969), pp. 4853–58.
Book:	[2] W. F. Hosford, *Mechanical Behavior of Materials*, Cambridge University Press, New York (2005).
Paper in a symposium or anthology:	[3] W. A. Backofen, "Formation of Slip-Band Cracks in Fatigue," in B. L. Averbach, D. K. Feldbeck, G. T.

	Hahn, and D. A. Thomas, eds., *Fracture*, Technology Press, Wiley, New York (1959), pp. 435–49.
Industrial report:	[4] J. C. Fister and J. F. Breedis, "Degradation and Recovery of Damping in Incramute," *Final Report, INCRA Project No. 274*, International Copper Research Association Inc., New York (1978).
Thesis:	[5] L. W. Leary, "Damping Degradation in Incramute and Sonoston Due to Low Temperature Storage," *Master's Thesis in Engineering Science*, Naval Post Graduate School, Monterey, Calif. (1986).
Private communication:	[6] D. C. Van Aken, Missouri University of Science and Technology, Rolla, Mo., personal communication (January 1, 2008).
Unknown author:	[7] *Making, Shaping and Treating of Steel*, 9th ed. United States Steel Corp. (1961), p. 1176.

Specific page numbers are usually cited unless the reference is to the entire book. Book and journal titles should be underlined or in italics. Some publishers and journals do not require that the title of the article be included. A reference may be cited several times using the same number, but it should appear only once in the reference section.

The Latin abbreviation *ibid.* (*ibidem* – in the same place) may be used when an information source is used in subsequent references, provided there are no intervening references cited. In other words, if two or more consecutive references are from the same source, *ibid.* would be used. Note that *ibid.* is a common-enough occurrence in scholarly writing that it is not usually written in italics.

A system of referencing used in many British journals uses the author's name (or authors' names) with year of publication either directly cited or in parentheses. In this case, references are listed alphabetically by author's last name in the reference section.

For example, the citations in the text might be in one of the following formats:

> ... and to tension or torsion in the other (TAYLOR and QUINNEY 1931; SCHMIDT 1932)

> ... however, it is extremely difficult to check as to whether this is so, as PUGH has recently recognized (1953)

> ... as modeled by AVRAMI (1939, 1941)

> ... as in CAHN (1956a) and CAHN (1956b)

Note that reference to more than one paper by the same author(s) in the same year is handled by adding letters to their citations.

In the reference section, the citations are listed alphabetically:

AVRAMI, M. (1939), *J. Chem. Phys.,* vol. 7, 1103.
AVRAMI, M. (1940), ibid., vol. 8, 212.
CAHN, J. W. (1956a), *Acta Metall.,* vol. 4, 449.
CAHN, J. W. (1956b), ibid., vol. 4, 572.
PUGH, H. LL. D. (1953), *J. Mech. Phys. Solids,* vol. 1, 284.
SCHMIDT, R. (1932), *Ingenieur-Archiv,* vol. 3, 215.
TAYLOR, G. I., and QUINNEY, H. (1931), *Phil. Trans. Roy. Soc. A,* vol. 230, 323.

Some journals may prefer references by the same author to be listed in one item:

AVRAMI, M. (1939), *J. Chem. Phys.,* vol. 7, 1103; (1940) ibid., vol. 8, 212.
CAHN, J. W. (1956a), *Acta Metall.,* vol. 4, 449; (1956b) ibid., vol. 4, 572.
PUGH, H. LL. D. (1953), *J. Mech. Phys. Solids,* vol. 1, 284.

Appendices

Details of calculations, derivations of equations, or documentation of computer codes that are not essential (but are still valuable) to the presentation of the report can be placed in appendices. Often, detailed derivations can slow the reader and detract from the essential findings of the investigation. All appendices should be referred to in the text, for example, *A complete listing of the computer code is presented in Appendix I.* Appendices should be sequentially numbered (often using Roman numerals) in order of appearance in the text. The appendix or appendices usually appear after the reference section at the end of the report.

Table of Contents

Normally journal papers and technical reports shorter than 50 pages do not need a table of contents. The reader can find specific areas by looking at the headings.

Chapter Summary

The basic guidelines presented in this chapter for format and content are useful for general report writing or for writing journal papers. Writing technical papers for journals can be a rewarding endeavor. The work will be read by others and may become a seminal reference for future authors. The next chapter presents a technical letter format for report writing.

3 Technical Letters

Technical letters are used for communicating scientific or engineering results that are limited in scope. The letters may describe a single experiment or investigation of which the results need to be rapidly communicated. Technical letters are a common form of communication for engineers or scientists in industry. Technical letters can also be used for undergraduate laboratory report writing in which a less formal presentation is appropriate. This chapter describes the organization and basic format of a technical letter. Two examples of letter reports are given at the end of the chapter.

Organization

Organization of the letter should begin with why the letter is being written, conclusions of the investigation, and what actions the recipient needs to address.

This first paragraph is sometimes called an action summary. The body of the letter should support the conclusions and recommended actions. The letter can be organized into three levels of presentation. At the first level, the first paragraph and the figures provide the necessary information to understand the conclusions and recommendations of the investigation. Figure captions must be informative and summarize the findings presented in the figures. At this level, a supervisor can ascertain with minimal reading the major findings of the investigation. The body of the letter should provide greater depth. There should be a summarizing paragraph at the end of the letter. Appendices, where calculations, derivations, and special test procedures are presented, constitute a third portion. Information in the appendices should be supplemental and referenced in the text. The report should be completely understandable without reading the appendices.

Format

Information concerning date, recipient, author, and subject of the letter should be specified at the beginning. In some instances the letter will be given a corporate report number. The following format is suggested.

Date: January 02, 2007
To: Recipient of letter, title
           Corporate address
From: Author's name, title
           Corporate address
Subject: Investigation of...

Corporate letterhead may be used for the first page. Subsequent pages should be numbered in sequential order on plain white paper.

Action Summary

The first paragraph should begin with an explanation of why the letter is being written and identify the origin of the request. Actionable items should be listed in this summary. An action summary should stand alone without citing tables, figures, or references in the text. It may be a simple statement of why a component failed or a summary of work requested by the recipient of the letter. This summary also establishes what the reader should expect to learn from the report and thereby helps the reader to critically evaluate the report.

Text

The writing style for technical letters is the same as described in Chapter 2 for technical papers.

As in technical papers, each table and figure should be numbered, referenced in the text, and appear in the order in which it is referenced. Tables and figures should appear at the end of short letters, since their inclusion in the text may be disruptive to the reader. Footnotes may be used for referencing previous work. Acknowledgments are not required.

Sometimes during the course of an investigation a new discovery unrelated to the main objective is made. This new information can be reported in an appendix. Unrelated significant findings should be reported in a separate letter; otherwise they may be obscured by the requested report.

Summary

The technical letter should end with a summary statement to provide closure. This summary may include unrelated discoveries made during the investigation.

Example Letter 1

This technical letter responds to a request for failure analysis of CDA 260 (cartridge brass) tubes that showed cracking shortly after they were formed into a 180° bend for a heat exchanger.

Date: April 01, 2007
To: Mary C. Haroney, Plant
             Metallurgist
             Brass Tube, Inc.
From: John H. Holliday, Materials
             Engineer
             Technical Center
Subject: Failure analysis of cracked
             CDA 260 tube assemblies

Four cracked CDA 260 copper tubes
were submitted for failure analysis.
Cracks were observed on the inside
radius of the 180° bends of the cool-
ing tube assembly. Stress corrosion cracking was determined to be
the failure mode. A combination of
residual stress and exposure to an
ammonia-based chemical is believed
to be responsible. A stress-relieving
heat treatment of one hour at 260°C
(500°F) is recommended.

Four exemplar tubes were received
for failure analysis from production
lot #1257. The tubes were manufac-
tured from the Copper Development
Association (CDA) alloy 260, which
is a copper alloy containing 30 weight

percent zinc. Circumferential cracks
were observed on the inner radius of
the bent tubes. Cracked tubes were
abrasively cut to expose the fracture
surfaces and then examined using a
scanning electron microscope. These
cracks originate at the surface and
extend approximately 400 µm into the
tube wall (see Figure 1). Figure 2

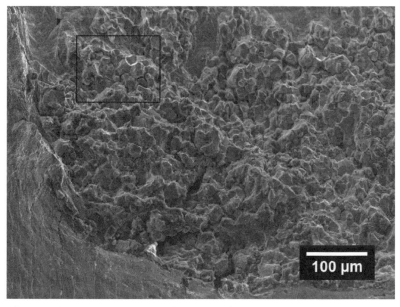

Figure 1. A secondary electron image of the
exposed crack showing an intergranular fracture, originating at the inside radius of the
tube bend. The intergranular cracks extend
400 µm into the tube wall. Figure 2 is located
by the box.

Figure 2. A secondary electron image of the
fracture surface that shows an intergranular
fracture mode. Intergranular fractures are
characterized by a "rock candy" morphology of
individual crystals or grains.

shows an intergranular fracture path
that is typical of stress corrosion
cracking of CDA 260.

Locations of the cracks relative
to the bent tubes indicate a tensile
residual stress resulting from elastic
spring-back after bending. The combination of residual stress and exposure
to ammonia-based chemicals is known

to cause stress corrosion cracking of CDA 260. It should be noted that this type of failure was first described as "season cracking," since it coincided with the spring application of fertilizers that are ammonia-based. A stress-relieving heat treatment of one hour at 260°C (500°F) directly after the tubes are bent is recommended.

In summary, analysis of the cracked CDA 260 tubes revealed a stress corrosion cracking failure mode resulting from residual stress and exposure to an ammonia-based chemical.

Example Letter 2

This technical letter responds to a quality check on a titanium alloy missile fin, which includes a chemical analysis and a quantitative measurement of a microstructural feature. This example includes a table, a figure, and an appendix showing relevant calculations.

```
Date:      April 01, 2007
To:        I. M. Indeep, Plant
           Metallurgist
           Ti-Cast of Missouri, Inc.
```

From: W. T. Pooh, Quality Control
           Department
           Ti-Cast of Missouri, Inc.
Subject: Chemistry and microstructure
           of Part No. 32283 — missile fin

The following is a report on
chemistry and microstructure of an
investment-cast Ti-6Al-4V missile
fin from production run #42357. The
chemistry was determined to be out
of specification, with higher-than-
acceptable levels of hydrogen and
chromium. The total residual alloy
content is within specification. The
customer will be requested to waive
the 1C7186 chemistry requirements
prior to shipping these parts. Past
practice of the customer has been to
accept parts that are slightly high on
hydrogen and chromium content provided
the total residual alloy content is
within specification. The microstructure is typical of the β-processed
Ti-6Al-4V missile fin with an α-plate
width of 3.25 ± 1.63 μm at a 95 percent confidence level (95% CL).

Chemical and metallographic analysis was performed on a part chosen

randomly from production run #42357.
A complete chemical analysis is shown
in Table I; the results indicate that
the missile fins are high in hydrogen
by 0.05 weight percent and residual
chromium by 0.03 weight percent. However, the total residual alloy content
of 0.26 weight percent is well within
the specified maximum of 0.4 weight
percent.

The missile fins had already been
annealed at 1070°C, which is approximately 60°C above the β-transus, and

Table I. *Chemical analysis in weight percent*

	Measured	2C7186 Specified
Titanium	89.96	remainder
Aluminum	5.4	5.5-6.75
Vanadium	4.1	3.5-4.5
Carbon	0.05	0.08 max
Hydrogen	0.02	0.015 max
Nitrogen	0.02	0.05 max
Oxygen	0.01	0.20 max
Residuals	(0.1 wt.% max each with total less than 0.4 wt.%)	
Chromium	0.13	
Molybdenum	0.05	
Iron	0.08	
Total residuals	0.26	

fan cooled. The microstructure consists of lamellar α+β colonies as
shown in Figure 1 and is typical of
our production parts. The prior β-
grain diameter was greater than 300 µm.
As requested, the α-plate width was

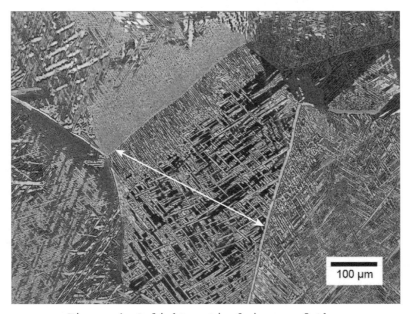

Figure 1. A light-optical image of the
annealed missile fin (part no. 32283). The
arrow in the figure shows a large (>300 µm
in diameter) prior β-grain structure that
has transformed to a mixture of α-plates and
retained β-phase. The microstructure is typical of a β-annealed Ti-6Al-4V alloy. The
microstructure was revealed using Keller's
reagent.

measured to be 3.25 ± 1.63 μm (95% CL)
using a mean linear intercept method.
Procedures and calculations for this
measurement are shown in the attached
appendix.

The measured hydrogen content is
more typical of the as-cast product.
Annealing will typically reduce the
hydrogen to acceptable levels provided the annealing is performed in
a slightly oxidizing environment.
It should be noted that the oxygen
content reported in Table I is 0.01
weight percent, which is low compared
to the more typical range of 0.1 to
0.15 weight percent for the annealed
product.

In summary, the missile fin parts
from production run #42357 should be
accepted provided the customer waives
the limitations on hydrogen content.
If the customer rejects these parts,
they may be salvaged by a second
annealing heat treatment. The Quality Control Department will immediately inspect the annealing furnaces
to determine if possible changes in
the furnace operation or personnel may

have affected the heat treatment of
the missile fins.

Appendix: Determination of the Mean Linear Size of Alpha Plates

The volume fraction of the β-phase was
determined to be 0.20 ± 0.04 (95% CL)
by a standard point-counting technique
using a 7 × 7 grid. The grid was placed
randomly on the microstructure (see
Figure A) ten times to obtain a confidence level (95% CL) that was less
than 20 percent of the average value.
A mean linear size of the α-phase
plate width was then determined to
be 3.25 ± 1.63 μm (95% CL) by placing
a line of length 130 μm on the photograph as shown in Figure A. Thirty-two
α-plates were intercepted, and the
mean linear size, L_3, was calculated
using

$$L_3 = \frac{\left(1 - V_f^\beta\right) L_{tot}}{N_{tot}^\alpha} = \frac{(1 - 0.2)\,130\,\mu m}{32} = 3.25\,\mu m,$$

where V_f^β is the volume fraction of
the β-phase, L_{tot} is the length of the
line on the photograph at the image

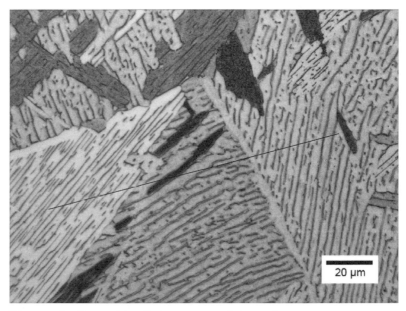

Figure A. An optical image used to calculate the mean linear size L_3 of the α-plates. The α-phase forms a continuous matrix, which etches with varying contrast based on the lamellar α-plate colony orientation. Thin ribbons of β-phase delineate the individual α-plates within the colony. Etching was performed with Keller's reagent.

magnification, and N^{α}_{tot} is the number of α-plates intercepted. A 95% CL for the mean linear size was calculated using the following:

$$\Delta L_3 (95\% CL) = \frac{4 L_3}{\sqrt{2 N^{\alpha}_{tot}}} = \frac{4 \left(3.25 \mu m\right)}{\sqrt{2 \left(32\right)}} = 1.63 \mu m.$$

Chapter Summary

In summary, letter report writing is the most commonly used form of communication. Most engineers and scientists spend their entire careers writing letter reports. This chapter presented the basics of organizing and formatting a letter report. The examples also showed the importance of presenting significant results in figures and tables. A more detailed discussion of data presentation is provided in Chapter 5.

4 Oral Presentations

At some point in the career of every scientist and engineer, they will be asked to present their findings to an audience of peers or a group of investors. In both cases, the purpose of the presentation is to sell ideas to the audience. Effective presentations maintain the attention of the audience. This chapter provides some basic guidelines for preparing an effective oral presentation.

Assessing the Audience

In planning an oral presentation, it is important to consider the knowledge level of the audience. Usually there is a wide discrepancy between backgrounds of different people. A talk should begin by telling the audience something they already know and then gradually work up to new material. A speaker should not try to impress the audience with his or her knowledge, since that will turn off most listeners. It is better to give a general talk that 95 percent of the audience can understand than an in-depth talk that only one or two people can comprehend. The number of equations in the presentation should be limited. Most audiences cannot digest more than three equations. Trying to cover too much material in the allotted time is a common mistake. A presentation should be pared down to one main theme.

Organization

Presentations must be carefully planned and organized to finish in time so that the audience can ask questions. Using too little time is preferable to using too much time. After giving a 30-minute talk, a noted scientist was asked how long it took him to prepare it. His answer was "About two days. Thank goodness it wasn't a 15-minute presentation. That would have taken a week to prepare."

Starting with a joke may be okay, but a poorly executed joke has been the downfall of many a speaker (including politicians), especially since a joke is not necessary. A talk should start by telling the audience what the presentation is about. A summary visual can be used to describe each part of the talk. The audience should be told when each section of the talk is finished so they can assess the progression.

The speaker should end with a summary of the main points of the talk (i.e., a summary of the conclusions).

Practice

The first step in preparing a talk is to write it out. This does not mean the speaker should read a prepared speech. But writing it out will help fix the main points in his or her mind. In addition, if the talk is read aloud as the visuals are reviewed, it will further fix in the speaker's mind what should be said and will allow the length of the talk to be measured. This is where it must be decided what should be omitted to stay within the time limit. If stage fright is a possibility, the first sentence or two can be memorized.

Speaking

The attention an audience gives to a speaker depends on the enthusiasm of the speaker. Enthusiasm is contagious. If the speaker exhibits great interest in his or her subject, the audience will think it is important too.

One should speak in full sentences, avoiding useless fillers like *umm...*, *err...*, *aah...*, *so...*, *all right...*, *I don't know...*, *I guess...*, *right?*, and, especially, *you know*. These utterances are distracting. However, it is often difficult to avoid them, and much practice may be necessary.

Redundant phrases like *very, very* or *very unique* should be avoided. Avoid the word *myself* when either *me* or *I* is meant. For example, *Professor Smith and I did the experiment* is correct, but *Professor Smith and myself did the experiment* is not. Similarly, *The experiment was done by Dr. Evans and me* is correct, but *The experiment was done by Dr. Evans and myself* is not.

Question and Answers

The speaker should allow sufficient time for questions after a presentation. The speaker should listen carefully to each question and have it clarified if it is not clear. The speaker should repeat the question so the entire audience knows what question is being answered. The answer should be direct. There is nothing wrong with saying *I don't know* in reply to a question. That is much better than guessing, unless it is made clear that the answer is a guess.

General Comments

The speaker should dress neatly. It is important to start the talk by greeting the audience, to look at the audience throughout the talk, and not to focus on the projection screen. One technique, called walking the triangle, has the speaker shifting both stage position and facial direction between the computer, the projection screen, and the audience. In this way, the speaker can change the visuals, verify that the audience can clearly see what is being presented, point out special features to be noted, and address the audience directly. The process of walking the triangle should be repeated for each visual used. Speak loudly enough for someone in the last row, who is hard of hearing, to hear. An associate can help monitor this. Also, blank slides during the presentation will break the monotony for the audience and focus their attention on the speaker. Be sure to acknowledge co-workers and help from others.

Visual Aids

In a talk, good visual aids are extremely important. The most common mistake made in creating a visual aid is putting too much on a single slide or viewgraph. With too much material, the size of everything is too small for people in the back of the audience to read what is written or discern what is plotted. Limit each visual aid to a single idea. Figure 4.1 shows a visual that is trying to show too much and so is illegible. In particular, the font size is too small and the underlining of the titles make them difficult to read. This material could be separated into three visuals: one for the advantages of magnesium alloys; a second

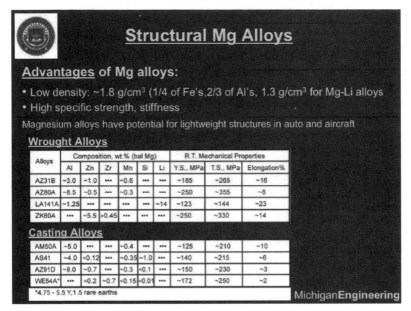

Figure 4.1. An example of too much material on one visual.

for the chemistry and properties of wrought alloys; and a third for the chemistry and properties of casting alloys. The table of wrought alloys could be shortened by omitting the column for silicon, and the table of casting alloys could omit the column for lithium. Furthermore, the tables could be widened to the full width of the overhead and *Michigan Engineering* at the bottom should be deleted.

Figure 4.2 is another example of too much material on a single slide. There are too many words in the bulleted items and the visual template occupies too much of the workspace. This can be improved by

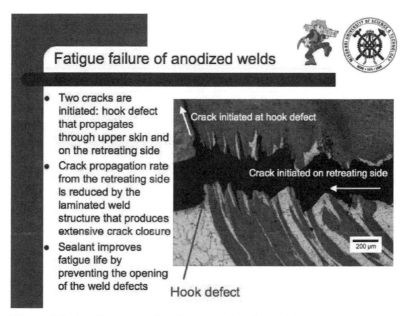

Figure 4.2. Another example of too much information on one visual.

having the photograph occupy the full width or height of the slide and having a separate slide with a few bulleted words. Larger type would also help, as would omitting the various university symbols.

General Principles for Computer-Generated Presentations

Preparing for the Presentation

Prior to a presentation, the speaker must be sure that the appropriate equipment and connections are

Fonts

Use

Arial
Geneva
Helvetica

Not

Brush Script MT
Impact
Lucidia Handwriting

Figure 4.3. Demonstration of font type and readability. Sans serif fonts, which use a simple stick construction, tend to be more readable from a distance. However, even these fonts are more difficult to read when in bold. Of course, Symbol font is useful for Greek letters in equations.

available. Otherwise, there may be a long and embarrassing delay while they are found and set up.

Font and Font Size

Avoid fancy fonts and long sentences; these are difficult to read when projected. Figure 4.3 shows fonts with simple stick construction (sans serif fonts) that are more readable than fancy fonts. Even sans serif fonts become less readable when bold, italic, or all

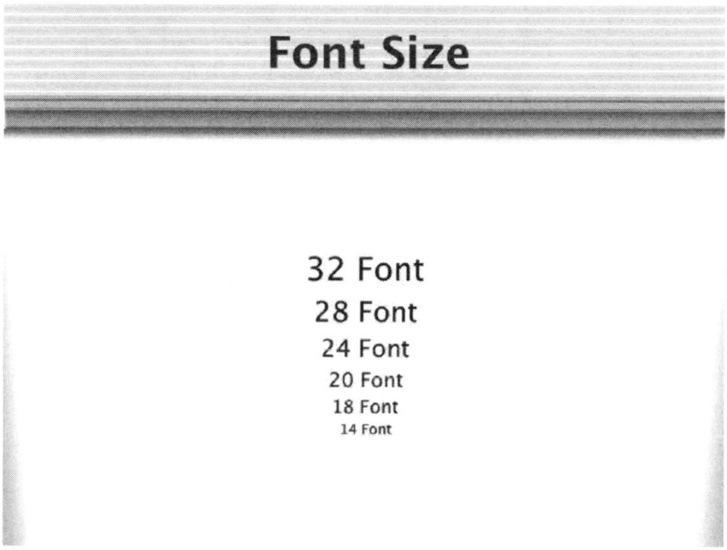

32 Font
28 Font
24 Font
20 Font
18 Font
14 Font

Figure 4.4. Demonstration of font size and readability. It is better to use a font that is at least 20 points.

capital letters are used. Figure 4.4 shows that readability is also affected by font size. Use a font larger than 24 points when the slide background is dark and the font is a light color; however, black 20-point type on a white background can be easily read. In general, a dark font on a light background is easier to read than a light font on a dark background.

Rules of Six

A bulleted line should contain no more than six words. Each slide should contain fewer than six

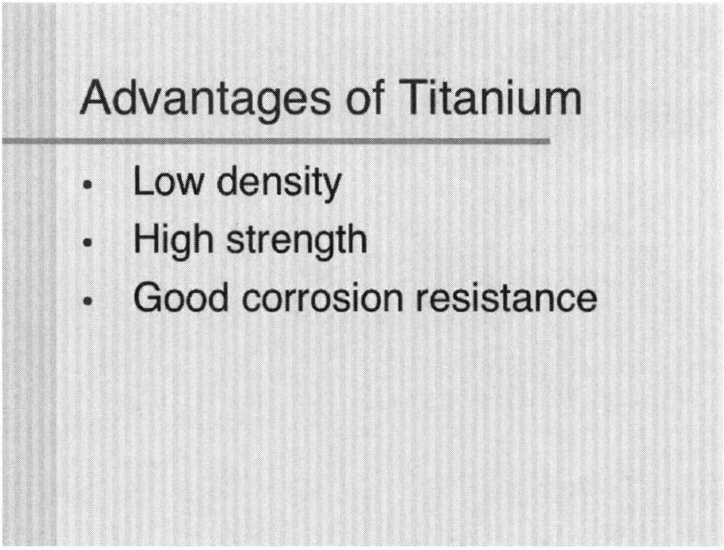

Figure 4.5. Appropriate use of bullets.

bullets, and no more than six slides containing only words or bulleted lists should be grouped together. Bullets are okay when listing points as shown in Figure 4.5. Each bullet should be used as a point of discussion in the presentation. An audience taking notes can quickly write down a few key words as a means to remember what was said. All too often, visual aids are used as comprehensive note cards that serve better as a textbook than as visual aids during a presentation.

Alignment

The text in slides should be justified left, centered, or justified right. One alignment should be chosen and used consistently throughout the presentation. Keep text and graphics away from the screen edges to prevent them from being obscured when the screen is not big enough for the projection system.

Colors and Special Effects

Use strong contrasting colors in all the graphics; it is likely that at least one person in the audience is color-blind. Backgrounds that transition from dark to light only serve to reduce the amount of usable area on the slide and, as a result, force the use of smaller type. The number of organizational affiliations that appear on each slide should be minimized; the audience only needs to read the affiliations once. Clip art should be used sparingly and animations avoided unless they demonstrate specific points in the presentation. Bouncing icons and sound effects only serve to distract the audience.

Chapter Summary

Simple presentations force the audience to focus on the speaker. Keep in mind that many of the great orators in history had nothing more than note cards from which to read. The goal of a presentation is to have the audience remember the ideas and invest in the technology. If the presentation is at a convention and a new job is of interest, remember that there are potential employers in the audience. In this case, it would be prudent to have a resumé handy. Chapter 7 describes some basic guidelines for preparing a resumé.

5 Presentation of Technical Data

This chapter covers some basic guidelines for presenting technical data. Tables, schematic drawings, photographs, and graphs quickly convey technical information and experimental results. A well-prepared table or figure immediately describes the significance of the work and provides a useful tool for the reader. Preparing tables and figures should be treated with the same care as writing a resumé. Technical reports are often the basis of patents, and the use of the standard international system of units (SI) is required for foreign and domestic patent applications. Appendix IV gives the international system of prefixes and units. Use of SI units is recommended, but the author should report in units and symbols most appropriate for the subject and the audience. Common uses of the Greek alphabet are provided in

Appendix V. Results can also be reported using multiple units. This chapter also provides guidance for preparing graphs with multiple scales in different units.

Tables

Tables are helpful for presenting and archiving experimental data. Unlike graphs, tables preserve exact numbers for future analysis. Tables should be sequentially numbered (often with Roman numerals) in the order of presentation in the text. Alphanumeric numbering is used in long reports that are broken into sections. Each table should have a simple title at the top. Tables should be incorporated into the body of the text as soon as convenient after they are referred to (they should not be placed in the middle of a paragraph unless the paragraph breaks across pages), or they can be grouped at the end of the report.

Tables make comparing data easy, so they should be constructed to simplify comparison. Numbers to be compared should be in adjacent columns or rows or both. The structure of Table 5.1 follows these guidelines: comparable heat treatments are in adjacent columns and the numbers to be compared are in the same row.

Table 5.1. *Mechanical properties of an aluminum 2219 alloy tested in tension*

	Test results for naturally aged (T4)	2219-T4*	Test results for artificially aged (T6)	2219-T6*
Young's modulus, GPa	71	71–73	72	71–73
Proportional limit, MPa	110		220	
Yield strength at 0.2% offset, MPa	136	185 min.	295	290 min.
Ultimate tensile strength, MPa	317	360 min.	426	415 min.
Percent elongation, 0.5-inch gauge	26	20 min.	16	10 min.

* Properties for 2219 are from Hatch (1984). Percent elongation was reported for a 2 inch gage length.

Figures and Figure Captions

Figures include any visual material – except tables – that may aid the reader. Figures should be numbered in the order in which they appear in the text; that is, the fifth figure referred to in the text should be Figure 5. All figures should be referred to in the text. Each figure should have a caption underneath it (not a title above it) that provides enough descriptive text to allow the figure and caption to stand alone and convey a significant finding. The caption, which may be several sentences long, should tell the reader what to look for in the figure. The number of figures should be limited to those necessary to justify the conclusions of the report.

Schematic Drawings

Schematic drawings should be labeled to indicate important parts. Do not expect the reader to know the features of the drawing. The same applies to features in photographs. Figures 5.1 and 5.2 are schematics of an electrical circuit and material flow in a chemical process.

Schematic drawings can also provide the basis of a theoretical analysis, as shown in Figure 5.3 for hardness testing, or show how a process works. Figure 5.4 shows schematically where a solid mandrel is placed in a tube while the tube is being bent. Note that the figure captions explain each of these drawings.

Equipment Photographs

Photographs of equipment or equipment models should clearly identify the important features. Figure 5.5 shows a conceptual model of a modern recoilless rifle. An alternative to labeling features in a photograph is shown in Figure 5.4, where each part is labeled with a letter and explained in the figure caption.

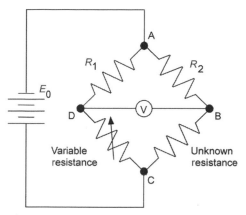

Figure 5.1. A Wheatstone bridge for measuring resistance with an applied voltage of E_0. The resistances R_1 and R_2 are equal. The variable resistance is adjusted until there is no voltage between D and B. The value of the unknown resistance equals the resistance of the variable resistor at that point.

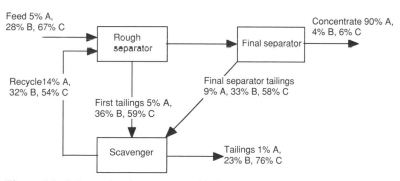

Figure 5.2. Schematic of a separator designed to enrich the concentration of A. Tailings from rough and final separators are recycled.

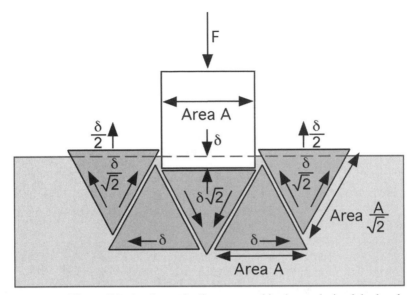

Figure 5.3. A schematic diagram used in the analysis of the hardness indentation. The five triangular portions slip relative to each other performing work under the applied load and displacement δ of the indenter. Figure adapted from Ashby and Jones (1981).

Photomicrographs

A scale bar in the photograph should be used to indicate magnification. Stating the magnification in the caption is not required or beneficial because the publisher may alter the size of the figure or the report may be reproduced at a different magnification. When the figure shows a microstructure, it is standard practice to include in the caption the chemical etchant used to reveal the structure. Each phase

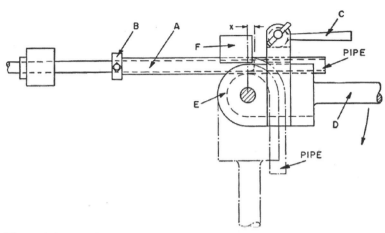

Figure 5.4. Drawing of a mandrel through a tube while it is being bent preserves the circular cross section. In the upper sketch, the tube to be bent is pushed over a mandrel (A) and against a stop (B), which locates the bend. The tube is clamped by a lever (C) and pulled by a lever (D), causing the form (E) to rotate (Schubert 1953).

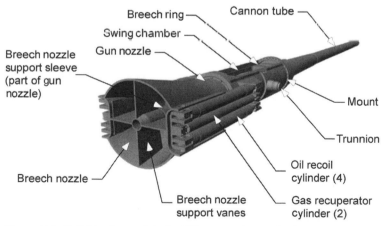

Figure 5.5. Solid-body model of a 105-mm sonic rarefaction wave gun or recoilless rifle (Kevin Miner, Benet Laboratories, September 15, 2006).

should be identified in both the figure and the caption (see Figure 5.6). One should not assume that the reader can identify the phases. However, some microstructures, such as the titanium microstructures shown in the second example letter of Chapter 3, do not lend themselves to labeling; such microstructures should be explained in the caption. When reports are reproduced, the significant features used to identify the phases may become obscured in the copies. Thus, clear labeling and descriptive captions are essential to the integrity of the report.

Graphing

Graphs are a good way of presenting data so the reader can see trends. Although most graphs are now produced using personal computers, there are some common pitfalls to avoid. Avoid the use of background colors or shading as they can obscure the data. Using color for data points should also be avoided. Most publications are in black and white, and grayscale renderings of shaded graphs or lightly colored data points are difficult to see. Do not include a title for the graph; that information should be in the caption. Enclosing the graph in a box wastes space. As a general rule, the graph should be legible after it has been converted to grayscale and reduced to

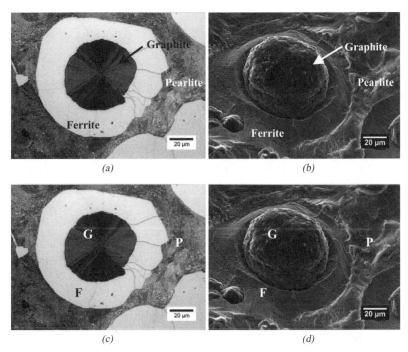

(a)   *(b)*

(c)   *(d)*

Figure 5.6. Two techniques for labeling features in an optical image of a nodular iron microstructure. In (a) and (b) the
microstructure is labeled directly and the reader need not read
the caption to interpret the image. However, large portions of
the image can be obstructed using this technique. Symbols (F-
ferrite; P-pearlite; G-graphite) are used in (c) and (d) to label
the microstructure. The images are of a ductile iron where the
graphite forms nodules. Polarized light was used to show the
radial growth of the graphite nodule. A 2 percent nital etchant
(2 volume percent nitric acid in ethanol) was used to reveal the
ferrite and pearlite microstructure in the optical images (a) and
(c). The nodule is encapsulated in ferrite (F) and the combination is often referred to as the bull's-eye microstructure. The ferrite grains are attacked along crystallographic planes when deep
etched (5 volume percent bromine in methanol), as revealed by
the secondary electron images shown in (b) and (d).

75 mm by 75 mm size. The following sections provide some basic guidelines on the science of graphing.

Ordinate vs. Abscissa

Normally, one plots the dependent variable on the ordinate (y-axis) and the independent variable on the abscissa (x-axis). For example, suppose the current through a rectifier is studied as a function of voltage; the current is the dependent variable and should be plotted on the ordinate. Another example is gasoline consumption by an airplane with respect to speed. In this case, gallons per mile is the dependent variable and velocity is the independent variable. Sometimes there is confusion on this point in stress-strain curves. Conventionally, stress is represented as the dependent variable (y-axis) and strain as the independent variable. This corresponds to the way in which most tension tests are made. The testing machine forces an elongation, and we measure the resulting force, which depends on the test bar. If one were to make a tension test by dead weight loading (e.g., adding a known weight of sand to a bucket hung from the test bar), the load (stress) would be the independent variable. However, in this case, plotting stress on the ordinate is still preferred because this is the conventional way of representing stress-strain curves, and therefore it would be most easily interpreted by readers.

There are other occasions when this convention is not followed. For example, fatigue data are usually presented in the form of an *S-N* plot with the stress, *S*, as the ordinate and cycles to failure, *N*, as the abscissa. Clearly, stress is the independent variable and cycles to failure is the dependent variable.

There are also cases in which there is no clear-cut choice about which variable is independent and which is dependent. An example is a graphical correlation between the weights and wingspans of airplanes. Neither measure is more independent than the other. The author's judgment should be used in such cases.

Choosing Scales

The first step in making a graph is to select the scales. In general, the scales should be selected so that the data cover a reasonably large fraction of the graph. This is illustrated in Figure 5.7.

The divisions on the graph should represent multiples of 1, 2, or 5×10^n units (but not 3, 12, etc.), as shown in Figure 5.8.

If the slope of a plotted line is important for interpretation, the scales should be adjusted so that

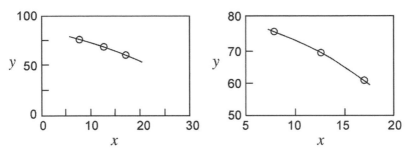

Figure 5.7. Choosing scales so the data occupy a large portion of the plot is important. The plot on the left could be expanded with an ordinate range of $50 \leq y \leq 80$ instead of $0 \leq y \leq 100$ and an abscissa range of $5 \leq x \leq 20$ instead of $0 \leq x \leq 30$, as shown on the right.

the slope is in the range of 30 to 60 degrees. It is very difficult for the reader to check a slope that is nearly horizontal or nearly vertical.

Whether the origin should be shown depends on several factors. One is the nature of the quantity being plotted. For example, if temperatures are being plotted in Fahrenheit, $0°$ has no special significance so there is no compelling reason to start the scale at

okay	0	5	10	15	20
or	0	10	20	30	40
or	0	0.02	0.04	0.06	0.08
but not	0	3	6	9	12
or	0	13	26	39	52

Figure 5.8. Examples of appropriate and inappropriate scale divisions.

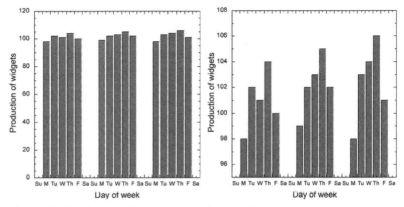

Figure 5.9. Inclusion of zero on a scale can obscure the importance of trends, as shown in the figure on the left. The figure on the right makes the differences appear greater and clearly shows that fewer widgets are produced on Monday than on any other day, that production steadily increases during the week, and that production decreases on Friday.

zero. The same is true of sound in decibels because there is no special significance to zero.

A second factor to be considered is whether one can show the origin and still have large enough divisions to show important variations. Consider a plot of how daily widget production varies through a three-week period (see Figure 5.9) when the daily production varies by less than 8 percent. Inclusion of zero on the *y*-axis makes the variation in widget production more difficult to see.

A third consideration is whether the data are being tested against or compared with a theory in which the origin has a special significance. Suppose the thickness of a chemical reaction product, L, has been measured as a function of time, t, and one wishes to compare the data with a theory that predicts that L is proportional to $\sqrt{t}$ (this means that $L \to 0$ as $t \to 0$). In this case, it would be appropriate to plot L vs. $\sqrt{t}$ on scales that do include the origin so one can see whether a straight line through the data extrapolates to the origin.

Labeling Scales

It is not necessary to indicate the scale level at every line. If too many lines are labeled, the graph will look cluttered; if too few are labeled, the scale will be difficult for the reader to interpret. A reasonable compromise is to indicate the scale level every 2 to 5 places, as shown in Figure 5.10.

It should be easy to tell which numbers on a scale refer to a major division. This can be a problem when the numbers are very large or very small, as indicated in Figure 5.11. In this case it is better to either plot 0.0005 as 0.5 on a $1000/T$ scale or to label it as 5×10^{-4}. Avoid ambiguity when labeling scales. For example, $1000/T$ indicates that the reciprocal of the temperature has been multiplied by 1000. This can also be written as $1/T \times 10^3$, but $1/T \, (10^3)$ is ambiguous and should not be used.

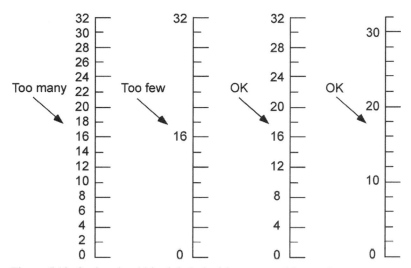

Figure 5.10. Scales should be labeled with a reasonable number of divisions. Too many labels make the scale difficult to read, whereas too few requires the reader to determine the values.

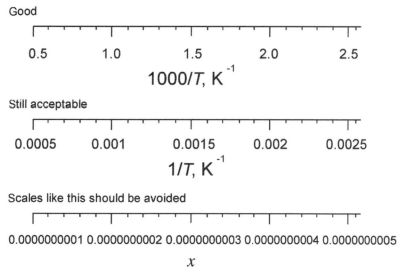

Figure 5.11. Examples of the appropriate use of numbers on axes to avoid overcrowding and a scale that is too crowded and difficult to read.

Use a font size of at least 14 points for the numbers in the scale and at least 16 points to label the axes. These guidelines will help retain readability when the graph is reduced for publication. A smaller font may be justified to avoid overcrowding. In Figure 5.9, 12-point type was used to avoid overcrowding the x-axis scale. However, expanding the x-axis and stacking the figures vertically rather than horizontally would avoid the overcrowding and permit the use of larger type.

Multiple Units on a Single Scale

Sometimes multiple units are used to describe the data in a graph. For example, stress data are often represented in both SI (MPa) and English (psi) units. If the left ordinate is labeled with SI units, the right ordinate can be in English units, or vice versa. The scale on the left ordinate can be indicated with tick marks at intervals of 1, 2, or 5×10^n. Intervals on the right ordinate should not coincide with intervals on the left ordinate; if they do, the numbers will not be simple (see Figure 5.12).

The same concept applies to labeling the axes to show both weight percent and atomic percent, both engineering strain and true strain, and the like. Remember, the axes must be labeled clearly to indicate both the variable being plotted and its units.

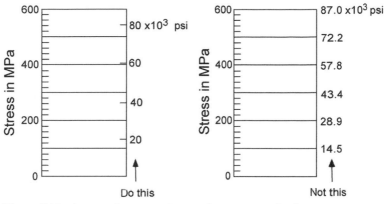

Figure 5.12. Appropriate use of more than one set of units.

Points

Experimental points should be plotted and appear large enough that they will be clearly visible even if the figure is reduced for publication. Calculated points are not generally indicated when a curve is theoretical. Error limits on the points may be included in the plot; the caption should indicate the level of precision or the amount of uncertainty that the limits represent.

Multiple Plots

Frequently, it is advantageous to plot more than one curve on the same graph. This saves paper and publication space (space costs money!). More importantly, it allows the reader to compare the curves.

When two or more curves are plotted on the same axes, it is essential that the reader be able to easily tell which curve is which and which points belong to each curve. If the curves are well separated and all of the points lie close to the curves drawn through them, it is sufficient to simply label each curve. However, it may be necessary to use different symbols for points on different curves ($\triangle$, ■, ◪, ○, □, ◇, etc.) or different types of lines for different curves (-----, ——, ... ,-·-·- -·-·-, etc.). In these cases, it is necessary to include a legend explaining the points and lines. An example of using different plot symbols and lines is shown in Figure 5.13 for the fatigue life of test specimens machined from titanium sheet by water-jet cutting and specimens that were subsequently polished to remove the damage created by water-jet cutting.

Drawing Curves Through Experimental Points

Should curves be drawn through every point or should they be smoothed to follow the general trend? Or should the data be approximated by a straight line? The answers depend on the circumstances. If there is reason to believe that the data are of sufficient accuracy that each hump and dip is real, then the curve should be drawn through all the points.

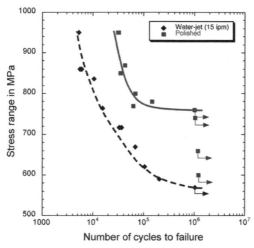

Figure 5.13. An example of two data sets being plotted in the same graph. The legend position should be chosen so that it does not obscure the data or interfere with reading the graph.

Consider Figure 5.14(a). Perhaps y is the temperature of a furnace and x is the time after it is turned on; the hills and valleys correspond to the cycling of the controller.

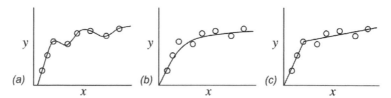

Figure 5.14. Three ways of drawing curves through experimental points: (a) curve drawn through all points, (b) smooth curve approximating points, and (c) data represented by two straight lines.

On the other hand, the same points can be represented by a smooth curve as shown in Figure 5.14(b), unless the deviation of such a curve from the points is greater than the possible error of the measurement or there is some compelling reason to believe that the phenomenon is cyclic. For example, y may represent the thickness of a growing film and x time. Finally, theory may play a role in how the data are best presented. Perhaps theory suggests that the behavior can best be represented by two straight lines, as in Figure 5.14(c).

Finding Slopes of Straight Lines

A straight line can be represented by the equation $y = ax + b$. The slope, a, can be found from two points, (x_1, y_1) and (x_2, y_2) as $a = (y_2 - y_1)/(x_2 - x_1)$. Two well-separated points should be used to avoid errors caused by experimental scatter (see Figure 5.15).

Grid Lines

The person making the graph should decide whether grid lines would be helpful to the reader. Grid lines may be useful if a reader needs to read specific values from the graph. They should be omitted if they obscure data or trend lines. A good compromise is to

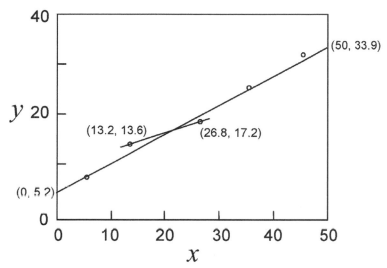

Figure 5.15. Determining the slope from points too close to one another can lead to great error. Here the true slope is $(33.9 - 5.2)/(50 - 0) = 0.57$, not $(17.2 - 13.6)/(26.8 - 13.2) = 0.26$.

use tick marks on the right and top scales as well as on the ordinate and abscissa.

Logarithmic Scales

Logarithmic scales are often labeled only at intervals differing by factors of 10 with no intermediate grid lines. If x is plotted on a logarithmic scale, the distance between two values x_1 and x_2 depends on the ratio of x_2/x_1. The distance on the paper between 1 and 2 is the same as the distances between 2 and 4 and between 5 and 10. In reading values between

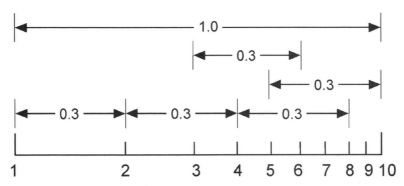

Figure 5.16. Reading a logarithmic scale. Note that the paper distance between two points that differ by a factor of 2 is close to 3/10 the length of the decade. A single decade is shown in the figure.

1 and 10, keep in mind that 2 is at a point about 0.3 times the distance between 1 and 10, so 5 is represented by a point about 0.7 of the distance between 1 and 10 (see Figure 5.16).

There are several reasons for using logarithmic scales. Often, the quantity being plotted varies by factors of 10, 100, 1000, or more over the range of interest and we want to be able to distinguish 8 from 10 as much as 800 from 1000. Other times, there are theoretical reasons for using logarithmic scales. In these cases there are two options: one is to plot the logarithm of the quantity directly on a Cartesian scale. In this case the scale should be labeled accordingly. The disadvantage of this approach is that it is difficult for the reader to discern the real value of the quantity. The other option is to use a logarithmic

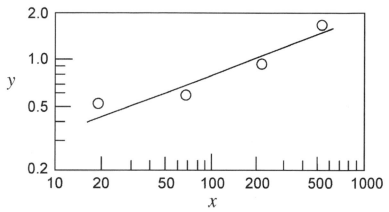

Figure 5.17 Labeling axes on a logarithmic scale. For the example shown, the additional labels on the y-axis are important, since only a partial decade appears on the y-axis.

scale and label it with the quantity directly. Often it is necessary to label the scales at a reasonable number of intervals (e.g. 0.02, 0.05, 0.10, 0.20, 0.50, 1.0, 2.0), as indicated in Figure 5.17.

Finding the Slope on a Log-Log Plot

If $y = ax^b$, then $\ln(y) = \ln(a) + \ln(x)$ and a plot of $\ln(y)$ vs. $\ln(x)$ on Cartesian coordinates or a plot of y vs. x on log-log paper will have a slope equal to b. The simplest way to find a slope is to take two well-separated points and realize

$$b = \frac{\ln(y_2) - \ln(y_1)}{\ln(x_2) - \ln(x_1)} = \frac{\ln(y_2/y_1)}{\ln(x_2/x_1)}.$$

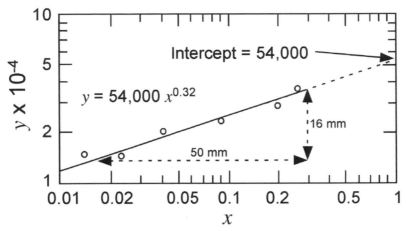

Figure 5.18. Finding the slope on a log-log plot. The slope may be determined by direct measurement using a scale where the rise and run are measured on the paper. In this example, the slope is equal to 16 mm/50 mm or 0.32.

Alternatively, the slope can be found by simply using a ruler to measure the x-distance and the y-distance and correcting them for the distance on the plot of a decade:

$$b = \frac{\Delta y \text{ in mm}/y \text{ decade in mm}}{\Delta x \text{ in mm}/x \text{ decade in mm}}.$$

Do not use the numbers on the scales of the log-log paper to determine the slope directly.

Note that if $y = ax^b$, a equals the value of y where $x = 1$ (see Figure 5.18).

Finding the Slope on a Semi-Log Plot

Arrhenius rate equations are often encountered in physical chemistry; the activation energy, Q, can be determined from a semi-log plot. The temperature dependence of diffusivity is a typical example of an Arrhenius-type relationship:

$$D = D_0 \exp \frac{-Q}{RT}.$$

A semi-log plot for the diffusivity of silver is shown in Figure 5.19. In this example the slope is equal to $-Q/R$, where R is the universal gas constant. Unlike the log-log determination for the slope, the actual values used to calculate the slope in a semi-log plot must be determined from the graph. For the exponential relationship the following formula can be used to calculate the slope:

$$slope = \frac{\ln(y_2/y_1)}{x_2 - x_1}.$$

In the example shown in Figure 5.19, the slope would be calculated as

$$slope = \frac{-Q}{R} = \frac{\ln\left(10^{-9}/4 \times 10^{-11}\right)}{(0.904 - 1.065)} \times 1000$$

$$= -19{,}990 \text{ K},$$

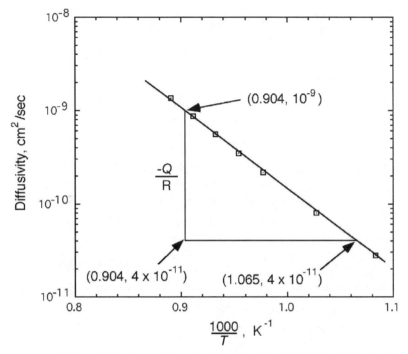

Figure 5.19. A semi-log plot showing the diffusivity of Ag in Ag as a function of the reciprocal temperature. The slope of the trend line is equal to $-Q/R$, where Q is an activation energy and R is the universal gas constant.

and the activation energy, Q, is determined as

$$Q = -R \times slope = -8.314 \frac{J}{mol \cdot K} \times -19{,}990\,K$$

$$= 166{,}200 \frac{J}{mol}.$$

Figures Generated by Computer Screen Prints

Computationally intensive computer programs are becoming an increasing part of the science and engineering professions. Finite element analysis for stress calculations and computational fluid dynamics for fluid flow and heat transfer are the most common examples. Figures showing calculated results are often generated as screen prints, but what appears to be readable on the computer screen is seldom legible when reduced for publication. An example is shown in Figure 5.20 for the calculated temperature profile of two aluminum bars that are 17.8 cm (7 inches) in diameter that are being heated by hot gas flowing from left to right. Figure 5.20(a) shows a typical screen print where the temperature scale is in scientific notation and the units of temperature and time are not explicitly defined. The same calculation is redrawn in Figure 5.20(b) to be more reader-friendly. The only information missing from Figure 5.20(b) is the time lapse for the calculation; this information should be included in the caption.

Chapter Summary

This chapter dealt with the visual presentation of data and best practices for graphing. It is often desirable to show a relationship between measured quantities and

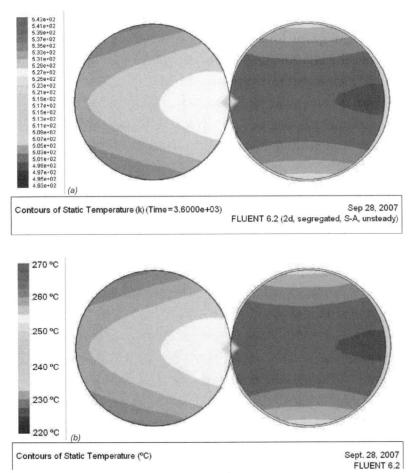

Contours of Static Temperature (k) (Time=3.6000e+03)   Sep 28, 2007
                                                FLUENT 6.2 (2d, segregated, S-A, unsteady)

Contours of Static Temperature (°C)   Sept. 28, 2007
                                                                        FLUENT 6.2

Figure 5.20. Screen prints generated from the computational fluid dynamics program FLUENT 6.2. The results show the temperature profile of two 17.8-cm-diameter aluminum cylinders after one hour of heating by a hot gas flowing from left to right. (a) An example of a screen print that is difficult to read because of the temperature scale format. In addition, the screen image does not specify the units of time, and the units of temperature should be capitalized, that is, K rather than k. (b) The same figure reformatted to make the temperature scale more readable.

physical relations that have mathematical formulas. Linear plots often require transforming the data; several examples were shown. Appendix VI shows methods of plotting common functions to obtain straight lines. Uncertainty in the measured data will influence the fitting of a straight line; Chapter 6 presents some basic concepts on statistics and uncertainty.

6 Statistical Analysis of Experimental Data

This chapter is an introduction to statistical analysis. Reporting of average values relative to a trend line does not convey the significance of a measurement. Calculation of uncertainty or confidence levels is required. This chapter discusses Gaussian and Weibull distributions, which are the two most commonly used in science and engineering. The chapter concludes with a discussion of uncertainty analysis where reported values and confidence levels depend on one or more measured quantities. This chapter is not meant to be a complete work on statistical analysis. It should, however, provide the necessary background for undergraduate students to include statistical analysis when writing lab reports. The professional may also find this chapter a useful resource when writing technical reports.

Errors and Calibration

Random and systematic are the two principal types of errors that occur during experimental measurements. *Random errors*, sometimes called accidental errors, may be introduced by variations in the instruments or by the person making the measurement. For random errors, positive and negative deviations occur with equal probability. If the measurements are biased toward either positive or negative deviations, then a *systematic error* may be present. Either the instrument or the person making the measurement can introduce both random and systematic errors. When random errors occur, a statistical analysis of the measurements is often used to determine the *precision* or uncertainty of the measurement. *Accuracy* of the measurement can only be determined by measuring a standard of known value. Standards can then be used to *calibrate* the measurements and eliminate systematic errors. When properly calibrated, the accuracy of the measurement is defined by the precision (or uncertainty) of the measurement.

Hardness testing of metal will be used as an illustration of accuracy and precision. The accuracy of a Rockwell hardness test is usually determined by measuring the hardness of a standard test block. Standard test blocks have a specified range of hardness, expressed as an average value ± an uncertainty,

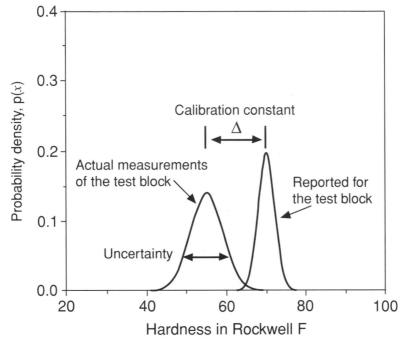

Figure 6.1. Hypothetical data representing measurements of a standard test block of known hardness. A systematic error is indicated by the difference between the actual measurements and the expected values. A calibration constant, Δ, is determined as the difference in the averages. The width of the distribution indicates an uncertainty in the measurement as a result of random errors.

and are used to calibrate the instrument. The uncertainty in the standard test block is a combination of the variation expected from the test machine and small variations in the material used to make the test block. Results from a hypothetical test are shown in Figure 6.1. In this example, the hardness of the test block falls below the expected results, indicating a systematic error and poor accuracy. The accuracy of the test may be improved by adding a calibration constant, Δ, that is calculated as the difference in the averages of the expected and actual results. After calibration, the accuracy is defined by the uncertainty of the actual measurements. It should also be noted that the uncertainty of the actual measurements is greater than that reported for the test block, and this may indicate random errors generated by the operator or that the hardness tester is in need of service.

Reporting Measurements

Results from an experiment are usually reported as the average of several individual tests. The arithmetic mean, or average, $\bar{x}$ is defined as

$$\bar{x} = \sum_{i=1}^{N} \frac{x_i}{N}, \tag{1}$$

where N is the number of measurements and the values of x_i are the individual values of each measurement. A *median* value can also be determined for the data set x_i. If the data are ordered in increasing value, the median is defined as the middle value for an odd number of results or as the average of the two middle values for an even number of results. A normal, or Gaussian, distribution will produce a median equal to the average.

Significant Figures

In reporting experimental results, the level of confidence is conveyed by the number of significant figures used to report the data. Using more significant figures than is justified by the actual measurements or precision of the measuring system will lead to an erroneous impression of the accuracy. Thus, the number of significant figures should be no greater than that used in measuring the original data.

It is also important to avoid ambiguity in the number of significant figures. Scientific notation can effectively be used to avoid possible confusion. For example, if a number is written as 33,500, the number of significant figures could be three, four, or five. However, using scientific notation and writing the number as 3.350×10^4 indicates that there are four significant figures and that the true value is between 33,495 and 33,504.

Experimental Uncertainty

With the exception of counting individual objects, all measurements are subject to accidental errors. The uncertainty in the reported average must be conveyed if the results are to have any significance. This uncertainty should be expressed as a $\pm$ value and the confidence level should be included for this uncertainty range, for example, as $\bar{x} \pm \Delta x (\%CL)$. The *confidence level* specifies the probability that the next measurement will fall within the specified uncertainty. For example, a 95 percent confidence level indicates that 95 of the next 100 measurements will fall within the uncertainty range, Δx, of the average, $\bar{x}$. Statistical analysis is required to establish both an uncertainty range and the confidence level.

Statistical Analysis of Experimental Data

A very simple method of describing the uncertainty is to specify the $\pm$ range as the average of the extreme positive and negative deviations about the arithmetic mean. However, this description depends on the number of measurements. With just a few measurements, there is a strong possibility that the next measurement will fall outside this specified range. There is no statistical method to calculate these odds. Thus, the uncertainty should define the frequency distribution, or dispersion, of the data.

Standard Deviation

The most common measure of dispersion in experimental data is the *standard deviation*, σ. Sometimes called the *biased standard deviation*, it applies to a

large number of individual tests and is calculated
thus:

$$\sigma = \sqrt{\sum_{i=1}^{N} \frac{(x_i - \overline{x})^2}{N}}. \qquad (2)$$

If the number of measurements, N, is small (fewer
than 20), then Bessel's approximation for σ can be
used and the result is an unbiased or sample standard
deviation, s, given by

$$\sigma \cong s = \sqrt{\sum_{i=1}^{N} \frac{(x_i - \overline{x})^2}{N - 1}}. \qquad (3)$$

It is sometimes useful to define a standard error for
the mean value, $\sigma_{\overline{x}}$. If the dispersion of the individual
measurements, σ_p, is known, then the standard error
for the mean value, $\sigma_{\overline{x}}$, may be calculated as

$$\sigma_{\overline{x}} = \frac{\sigma_p}{\sqrt{N}}. \qquad (4)$$

Typically, σ_p is unknown because many measurements are required to determine if the population is
Gaussian or is described by another statistical distribution, such as Weibull statistics. An estimate for $\sigma_{\overline{x}}$
may be obtained using σ, the standard deviation.

The dependence of yield strength on the grain
size in metals is an example where the calculation

of $\sigma_{\bar{x}}$ for the mean grain diameter would be useful. Measurements of the grain diameter are usually conducted on a single metallographic specimen where multiple grains are measured. The standard deviation of the grain diameter is representative of the variation in the individual grain diameters. In contrast, a tensile test will measure the yield behavior of a large number of grains. In a typical gage section (50.8 mm length by 12.8 mm diameter), there will be approximately 4.1×10^8 separate grains, assuming a mean diameter of 40 μm. If several specimens are tested, then the dispersion of the yield strength is representative of the dispersion of the arithmetic mean of the grain diameter and not of the dispersion of the individual grain diameters. Estimations of $\sigma_{\bar{x}}$ for developing microstructure-property relationships can be made using Equation (4) and the standard deviation of the individual grain diameters.

Confidence Levels

In order to specify a confidence level for a given uncertainty range, the exact probability density, $p(x)$, for the measurements must be determined. This information is not always available, or the number of measurements required to establish $p(x)$ may be too expensive to perform. As a result, a *Gaussian*, or *normal*, distribution is often assumed. The probability

Table 6.1. *Confidence levels based on the standard deviation and a normal distribution*

Uncertainty range, Δx	Odds of the next result falling in this range	Confidence level (%)
$\pm 0.6745\,\sigma$	1:1	50
$\pm\,\sigma$	2.15:1	68.27
$\pm 2\,\sigma$	21:1	95.45
$\pm 3\,\sigma$	369:1	99.73

density, $p(x)$, for the normal distribution is given by

$$p(x) = \frac{1}{\sigma\sqrt{2\pi}} \exp \frac{-(x - \bar{x})^2}{2\sigma^2}. \qquad (5)$$

A confidence level can now be established for an uncertainty based on the standard deviation and the normal distribution (Table 6.1). Other methods and statistics will yield different confidence levels; Table 6.1 only applies to a normally distributed population where N is large.

For small N it is inappropriate to set the uncertainty range at the 95 percent confidence level as simply $\pm 2\sigma$. A better method of calculating the uncertainty range is provided by

$$\Delta x(95\% CL) = ts. \qquad (6)$$

The value of t varies with the number of measurements (Table 6.2), and s is the sample standard deviation defined by Equation (3). The standard error of

Table 6.2. *The t values for calculating the 95 percent confidence levels*

$N-1$	t	$N-1$	t
1	12.706	18	2.101
2	4.303	19	2.093
3	3.182	20	2.086
4	2.776	21	2.080
5	2.571	22	2.074
6	2.447	23	2.069
7	2.365	24	2.064
8	2.306	25	2.060
9	2.262	26	2.056
10	2.228	27	2.052
11	2.201	28	2.048
12	2.179	29	2.045
13	2.160	30	2.042
14	2.145	40	2.021
15	2.131	60	2.000
16	2.120	120	1.980
17	2.110	∞	1.960

Source: Dieter 1991.

the mean at a 95 percent confidence level, $\Delta \bar{x}$, is then given by

$$\Delta \bar{x}(95\% CL) = \frac{ts}{\sqrt{N-1}}. \qquad (7)$$

Test for the Normal Distribution

Positive and negative deviations will occur with equal probability only if the distribution is normal. A convenient way of determining whether a set of data

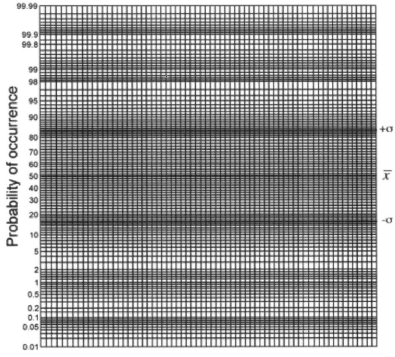

Figure 6.2. Normal probability paper. The standard deviation is determined as the difference between the arithmetic mean, $\Delta \bar{x}$, and the value at $+\sigma$ or $-\sigma$.

exhibits a normal distribution is to plot the cumulative frequency on normal probability paper, which is shown in Figure 6.2. The ordinate of the graph paper is distorted in such a manner as to produce a straight-line plot when the data have a normal distribution. To plot N measurements, the data are sorted in increasing order and given a rank, i, starting with the number one. An accumulative probability, or probability of occurrence, $P(x_i)$, is then calculated using

$$p(x_i) = \frac{i}{N+1} \times 100\%. \qquad (8)$$

The data are plotted on the abscissa, which is a linear scale. If the data produce a straight line, then the population is normally distributed. Values for the arithmetic mean and the standard deviation may be estimated directly from the plot. The arithmetic mean, $\bar{x}$, will be the value when the probability of occurrence is 50 percent, and one standard deviation above and one below the mean occur at probability levels of 84.13 and 35.87 percent. Estimation lines for these values have been highlighted in Figure 6.2. In small data sets, deviations from linearity should be expected at the higher and lower ends of the curve because the estimation of the probability of occurrence using Equation (8) is less reliable at the extremes.

A better approximation for the probability of occurrence, $P(x_i)$, can be obtained using Benard's approximation for median ranks. Johnson (1951) has shown this to be useful in the Weibull analysis of fatigue lives. The approximation is given by

$$P(x_i) = \frac{i - 0.3}{N + 0.4} \times 100\%. \qquad (9)$$

Sometimes the data may not show a normal distribution, but a simple mathematical transformation of the measurements may produce a normally distributed population. Two commonly used transformations are $x' = \log x$ and $x' = x^{1/2}$, where plotting x' instead of x exhibits a normal distribution. Fatigue data will generally have a log-normal distribution, and the logarithmic transformation of the number of cycles to failure for multiple samples tested at the same applied stress will produce a straight-line graph on normal probability paper. Log-normal probability paper is also available, but is not included here.

Unfortunately, the normal probability plot cannot be constructed using a simple mathematical transformation of the data. However, many graphics software packages will generate normal probability plots, as shown in Figure 6.3 for the fracture strength of a carbon-fiber-reinforced epoxy composite. The data for this graph are shown in Table 6.3. The data may be considered normally distributed as a first approximation to the actual distribution. However, a slight negative curvature (concave down) in the central portion of the population suggests that the distribution is not normal. Also, the strong negative deviation at the low end suggests that the population may exhibit a limiting stress below which the probability of failure goes to zero. It should be noted that a normal distribution suggests that there is a finite probability of failure at zero stress for the composite. Of course, this does not make physical sense. The normal

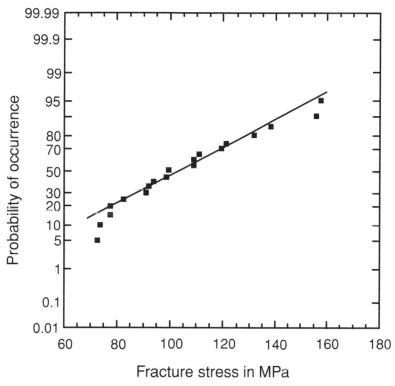

Figure 6.3. A normal probability plot of the fracture stress of a carbon-fiber-reinforced epoxy composite (see Table 6.3 for the data). This is an example where the data are not truly normally distributed. A slight negative curvature in the central portion of the population and deviations at the extreme ends indicate that the distribution is not normal. Figure 6.5 shows the same data assuming Weibull statistics.

distribution is such a powerful tool in the analysis of engineering data that slight deviations in the extreme portions of the population are often ignored to take advantage of its simplicity.

Table 6.3. *Fracture stress data for a carbon-fiber-reinforced epoxy composite*

Fracture stress, MPa	Rank	Probability of occurrence, $\frac{i}{N+1} \times 100\%$	$\frac{|x_i - \bar{x}|}{\sigma}$
72.5	1	5	1.28
73.5	2	10	1.24
77.7	3	15	1.08
78.2	4	20	1.06
82.7	5	25	0.889
90.8	6	30	0.580
92.4	7	35	0.519
93.7	8	40	0.469
98.9	9	45	0.271
99.5	10	50	0.248
109	11	55	0.115
109	12	60	0.115
111	13	65	0.191
120	14	70	0.534
121	15	75	0.573
132	16	80	0.992
139	17	85	1.26
156	18	90	1.91
158	19	95	1.98

The arithmetic mean $(\bar{x}) = 106$ MPa, and the sample standard deviation $(s) = 26.2$ MPa.

Chauvenet's Criterion for Discarding a Measurement

Even the best experimentalist may inadvertently produce data that appear questionable. For example, in Figure 6.3, a more linear plot might be obtained if the lowest measurement is removed. However, it is not appropriate to arbitrarily exclude measurements that do not meet expectations. Chauvenet's criterion (1989) provides a means to test the data and determine whether a particular measurement can be removed from a data set. It should be emphasized that this procedure allows only one measurement to be removed.

To apply Chauvenet's criterion, the arithmetic mean and the standard deviation are calculated for the data set in the usual manner. For small data sets the standard deviation (σ) can be approximated by the sample standard deviation (s). In addition, the ratio of the deviation, d_i, to the standard deviation, σ, is calculated for each measurement using Equation (10); these results are also shown in Table 6.3 for the fracture stress of the carbon-fiber composite:

$$\frac{d_i}{\sigma} = \frac{|x_i - \overline{x}|}{\sigma}. \tag{10}$$

Chauvenet's criterion requires that the ratio calculated using Equation (10) must exceed a specified value before the measurement can be excluded; this value depends on the number of tests, N (Table 6.4). Chauvenet's criterion assumes a normal distribution. According to Table 6.4, the maximum deviation for the group of 19 measurements is between 2.13 and 2.33. The largest deviation of the data in Table 6.3 is

Table 6.4. *Chauvenet's criterion for rejecting a measurement*

Number of measurements, N	Ratio of maximum deviation to standard deviation, d_{max}/σ
3	1.38
4	1.54
5	1.65
6	1.73
7	1.80
10	1.96
15	2.13
25	2.33
50	2.57
100	2.81
300	3.14
500	3.29
1,000	3.48

1.98, so all of the data must be included in the statistical analysis. If Chauvenet's criterion is met, then the arithmetic mean and standard deviation are recalculated after removing the dubious measurement. The value of N must also be reduced by one.

Weibull Statistics

Most physical properties exhibit a lower bound in the probability distribution, which the normal distribution fails to accurately describe. The Weibull distribution was originally proposed for describing fatigue life, but it has been used to model many different engineering properties, such as brittle fracture of ceramics and the life of electronic components. The probability density, $p(x)$, for the Weibull distribution is given by

$$p(x) = \frac{m}{\theta} \left(\frac{x}{\theta}\right)^{m-1} \exp\left[-\left(\frac{x}{\theta}\right)^{m}\right].$$ (11)

The shape of the distribution curve is controlled by the value of m and is referred to as the Weibull modulus. Example distributions, with varying values of m, are shown in Figure 6.4. The population distribution narrows rapidly as the value of m increases, and measurements with a high Weibull modulus are thought of as more reliable because there is less scatter in the data. The scaling parameter θ is called the characteristic value; at $x = \theta$ the population is divided into 63.2 percent below and 36.8 percent above θ for all values of m.

Calculation of a mean, Equation (12), and of a variance, Equation (13), for a Weibull distribution is not straightforward; these calculations involve the standard gamma function, Γ. However, the main reason for using Weibull statistics is not to report means or variances, but rather to report the probability of an event occurring.

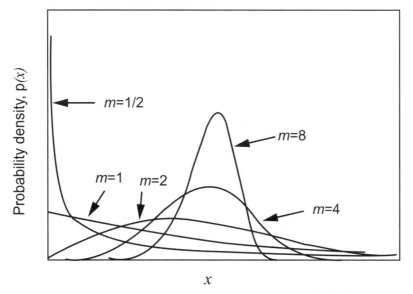

Figure 6.4. A schematic plot showing the Weibull distribution function with different values of m. In this plot $\theta = 1$ and $x_0 = 0$.

$$\bar{x} = \theta\Gamma\left(1 + \frac{1}{m}\right). \tag{12}$$

$$\sigma_p^2 = \theta^2\left[\Gamma\left(1 + \frac{2}{m}\right) - \Gamma^2\left(1 + \frac{1}{m}\right)\right]. \tag{13}$$

The probability of occurrence, $P(x)$, for the Weibull distribution is easily plotted with any graphics software package and the important parameters obtained graphically or by linear regression analysis. To incorporate a lower bound to the population, a third parameter, x_0, may be introduced. The three-parameter equation for $P(x)$ is given by

$$P(x) = 1 - \exp\left[-\left(\frac{x - x_0}{\theta - x_0}\right)^m\right]. \tag{14}$$

The probability of seeing a value less than x_0 is zero. Setting x_0 equal to zero in Equation (14) produces a standard two-parameter Weibull equation with a probability distribution characterized by Equation (11).

To produce a straight-line plot, Equation (14) is rewritten as

$$\log\left[\ln\left(\frac{1}{1 - P(x)}\right)\right] = m\log(x - x_0) - m\log(\theta - x_0)$$

or

$$\ln\left[\ln\left(\frac{1}{1 - P(x)}\right)\right] = m\ln(x - x_0) - m\ln(\theta - x_0).$$

Values of $P(x)$ are obtained in the same manner as with normal probability paper. The data are first sorted in ascending order and ranked. $P(x)$ is then calculated using Equation (8) or (9), with the exception that a fractional number is used rather than a percentage. The data are then plotted as

$$\ln\left(\frac{1}{1 - P(x)}\right) \text{ vs.}(x - x_0)$$

on log-log axes or on linear scales as

$$\log\left[\ln\left(\frac{1}{1 - P(x)}\right)\right] \text{ vs. } \log(x - x_0)$$

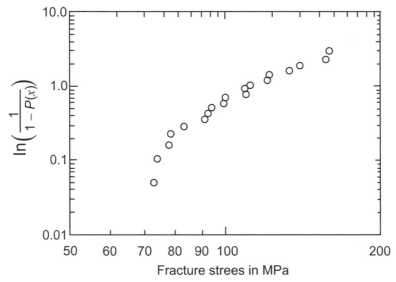

Figure 6.5. A two-parameter Weibull plot of fracture stress for the carbon-fiber-composite data shown in Table 6.3. A first approximation of x_0 for the three-parameter Weibull plot can be found by extrapolating an imaginary curve through the data and down to the abscissa. A value of 70 MPa is found using this method.

or

$$\ln\left[\ln\left(\frac{1}{1 - P(x)}\right)\right] \text{ vs. } \ln(x - x_0).$$

A two-parameter plot is shown in Figure 6.5 for the carbon-fiber-composite data in Table 6.3. The negative curvature indicates a value greater than zero for x_0. A first approximation of $x_0 = 70$ MPa is found by extrapolating an imaginary curve through the data and down to the abscissa. The best value of x_0 is found

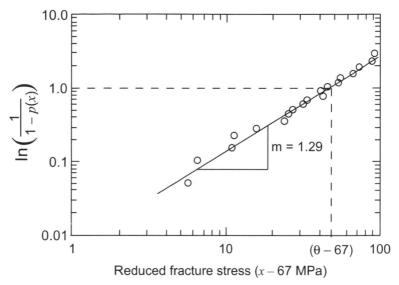

Figure 6.6. A three-parameter Weibull plot of fracture stress for the carbon-fiber-composite data shown in Table 6.3. A value of $x_0 = 67$ MPa was found to produce the best straight line.

by adjusting x_0 and observing the change in curvature. If the data show positive curvature, then the x_0 value is too high. The best straight line for the carbon-fiber-composite data was obtained by setting x_0 equal to 67 MPa; see Figure 6.6. Values for m and θ can then be obtained graphically, as shown in Chapter 5, or more directly using a linear fitting routine when the data are plotted as shown in Figure 6.7.

Extracting engineering information directly from the graph is a little easier if log-log scales are used as shown in Figure 6.6, but Figure 6.7 has the

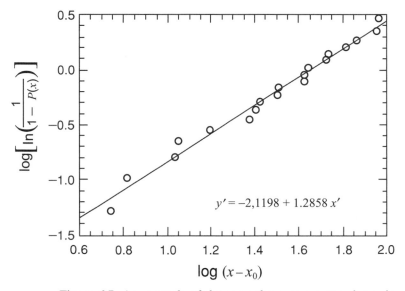

Figure 6.7. An example of the same three-parameter plot as in Figure 6.6, but using a linear scale on the ordinate and abscissa. A linear fitting routine can now be used to determine the equation of the best-fit line.

advantage of yielding the best-fit line from which values of m and θ may be calculated from the following relations:

$$y' = \log\left[\ln\left(\frac{1}{1-P(x)}\right)\right]$$

$$1.2858x' = m\log(x-x_0)$$

$$-2.1198 = m\log(\theta-x_0).$$

where $m = 1.2858$.

Probability of Failure Calculations

The probability of occurrence, $P(x)$, may also be thought of as the probability of failure when x represents a failure stress or the number of cycles to failure. Weibull analysis then provides information about the probability of failure that can be used in design. For the example of the carbon-fiber composites, a safe loading limit might be specified as 67 MPa because the probability of failure at this stress is zero. If the composite is used in a non–life-threatening application, then perhaps a failure rate of one out of a thousand is acceptable. Using the best-fit equation from Figure 6.7, an applied stress of 67.2 MPa would fail one out of a thousand $(P(x) = 0.001)$ carbon-fiber composites.

Example calculations:

$$\log\left[\ln\left(\frac{1}{1 - 0.001}\right)\right] = -2.99978$$

$$-2.99978 = 1.2858\log(x - 67 \text{ MPa}) - 2.1198$$

$$x = 67.2 \text{ MPa}.$$

Weibull analysis might also be used to predict the number of hours that a circuit can operate with only 0.0001 percent chance of failure. Instead of testing a million circuits, one can test 100 circuits and get reasonable values of m and θ to solve for x with $P(x) = 10^{-6}$.

Uncertainty Analysis

It is sometimes necessary to transform experimental data and the corresponding uncertainty by a mathematical operation to obtain a desired engineering result. A typical example would be the measurement of a stress. During a tensile test the load, F, rather than the stress, S, is actually measured and must be converted to a stress by dividing by the cross-sectional area, πr^2. Uncertainty in the stress value is introduced as a result of errors in measuring the sample radius, r, and errors in measurement of the load. Converting these errors to an uncertainty in the stress can be accomplished in a number of ways. A commonsense approach is to combine all of the errors in the most detrimental way to determine the minimum and maximum values that might be obtained. The following is an example of this approach for calculating the stress:

$$\frac{\overline{F} - \Delta F}{\pi (\overline{r} + \Delta r)^2} \le S \le \frac{\overline{F} + \Delta F}{\pi (\overline{r} - \Delta r)^2}.$$

A more precise method for calculating uncertainties was developed by Kline and McClintock (1953). To illustrate how this method was formulated, consider the relationship between y and x when they are related by the equation $y = \ln x$; see Figure 6.8.

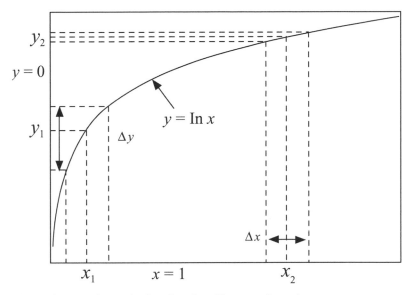

Figure 6.8. A schematic drawing that illustrates how the uncertainty in the x variable may be converted to an uncertainty in the y variable via a mathematical transformation. In this case y is related to x through the equation $y = \ln x$.

The Δy uncertainty will depend on the value of x, through the calculation of the slope at x, and Δx. For this particular example, it should be noted that small values of x produce large Δy uncertainties:

$$y = \ln x$$

$$\frac{dy}{dx} = \frac{1}{x} = \frac{\Delta y}{\Delta x}$$

$$\Delta y = \frac{\Delta x}{x}.$$

This method can be used to convert uncertainties in measurements when a mathematical transformation of the data is necessary to obtain a straight-line plot, for example, plotting $\ln x$ rather than x on a log scale.

When multiple variables $(x, y, z, \ldots)$ are used to calculate a quantity, w, and each has an uncertainty $(\Delta x, \Delta y, \Delta z, \ldots)$ associated with it, then the following general equation is used:

$$w = f(x, y, z, \ldots)$$

$$\Delta w = \sqrt{\left(\frac{\partial f}{\partial x}\Delta x\right)^2 + \left(\frac{\partial f}{\partial y}\Delta y\right)^2 + \left(\frac{\partial f}{\partial z}\Delta z\right)^2 + \ldots}$$

$$(15)$$

The resultant uncertainty will have the same confidence level as the uncertainties used in the calculation, provided they are all the same. Thus, if all the uncertainties are given at the 95 percent confidence level, then the result will also be at a 95 percent confidence level.

Example of Transforming Uncertainties

The hardness of recrystallized cartridge brass is dependent on grain diameter as described by

$$H = H_0 + \frac{k}{\sqrt{L_3}}, \qquad (16)$$

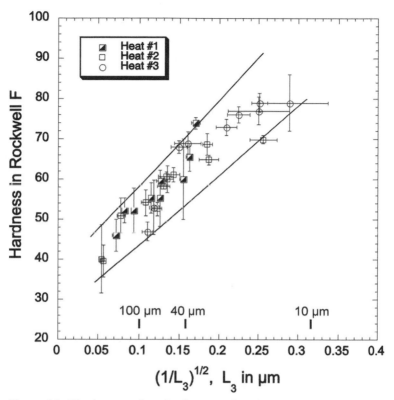

Figure 6.9. Hardness and grain diameter data for recrystallized cartridge brass. Data were collected from three consecutive productions, or heats, of cartridge brass. Uncertainties represent a 95 percent confidence level for the ordinate and abscissa values. A linear relationship is expected between hardness and the grain diameter based on Equation (16).

where L_3 is the mean linear intercept of the grain diameter, H_0 and k are materials constants, and H is the hardness.

A plot of H vs. $1/\sqrt{L_3}$ should produce a straight line (see Figure 6.9), and the uncertainty in L_3 would

be transformed to an uncertainty (Δx) on the abscissa
as follows:

$$x = \frac{1}{\sqrt{L_3}}$$

$$\Delta x = \sqrt{\left(\frac{dx}{dL_3}\Delta L_3\right)^2} = \frac{1}{2}(L_3)^{-\frac{3}{2}}\Delta L_3,$$

where ΔL_3 is the uncertainty in the measured value
of the grain diameter of L_3. Figure 6.9 is a compila-
tion of three consecutive productions, or heats, of the
cartridge brass, and the data follow an approximate
linear trend as shown by the upper and lower bounds.

Chapter Summary

Uncertainty analysis and reporting confidence levels can add credibility to a technical report; that is
why this chapter is included in this book. Although
this chapter provides basic guidance in using statistical analysis, it is not a complete treatment. However,
the information provided should be sufficient to treat
experimental results obtained in most undergraduate
science and engineering laboratory courses.

7 Resumé Writing

The purpose of a resumé is to obtain a job. Only a small fraction of resumés result in job interviews. The reader of resumés spends an average of 30 seconds on each one. To be successful, a resumé should be short, with the important information listed first. It should be well organized and neat. Often a resumé is tailored to a specific job, which would require rewriting the resumé for each new position.

Organization

The first step is to gather the pertinent information and organize it. Then this information should be divided into headings such as *Personal Information*, *Work Experience*, *Education*, *Skills*, *Honors*, and *Activities*.

Personal Information

The person's name, without titles, should appear at the top of the page in a larger font than the rest of the document; use this larger size type for the headings as well. Next list home address, phone numbers, email address, and fax number (if applicable). Citizenship may be listed, but this is not required. Personal information such as age, sex, and general health need not be listed either.

Work Experience

Experience includes full-time and part-time jobs, internships, academic research positions, and volunteer work. List the employer, with months and years worked, position title, and responsibilities. For example, *Sam's Café, September 2005 to June 2006, waiter* or *ABC Chemicals, June to August 2005, summer intern, preparing special orders.* Jobs should be listed in reverse chronological order, with the most recent first. Inventions and publications may be listed under Experience or separately under *Publications* or *Inventions* at the end of the resumé.

Education

Education listed on the resumé should include only college-level studies. The exception is if one is only in the first year of college and applying for a summer internship; then high school information may be included. Degrees, with month and year obtained or expected, should be listed, along with the name of the school, major (and minor if any), and grade-point average. Sometimes listing of important courses is appropriate.

Skills

Skills include facility in computer languages, foreign languages, teaching, communication, leadership, and teamwork. List the most important skills first.

Honors

Honors include scholarships, academic awards, and recognition of community service or athletic achievement. These are also listed in reverse chronological order.

Activities

Under Activities, list any student, professional, or community organizations and the various offices (e.g., president, treasurer, secretary) held in these organizations. Listing of extracurricular activities and hobbies is optional.

Wording

Certain words help create a favorable impression. Among these are:

ability	achieved	built	conceived	controlled
demonstrated	developed	devised	directed	enhanced
exhibited	expanded	generated	helped	imagination
improved	incorporated	installed	led	managed
motivated	organized	overcame	perfected	pioneered
produced	recognized	reduced	served	simplified
solved	streamlined	taught	unified	wrote

Their tense and voice may be changed.

A list of objectives can be added, but be careful not to be too vague or limiting.

Never lie or exaggerate; this could lead to trouble. For example, overstating a proficiency in a foreign language can lead to embarrassment during the interview if the interviewer asks a question in that language. Avoid humor and flamboyant wording; use a simple, easy-to-read font. There is no need for visual material.

Resumé Hints

1. Assume the reader is intelligent.
2. A resumé is not a curriculum vitae or autobiography; keep it short.

3. The grammar of resumés is simplified: Subjects of sentences and personal pronouns are usually omitted, for example, *As part of a three-man team, decreased rejection rate 5 percent.*
4. Avoid the negative. Instead of saying *Almost met the targeted reduction rejection rate of 4 percent,* say *Achieved a 5 percent rejection decrease.*
5. Showing accomplishments is not boasting. Saving "something" for the interview is a mistake. Significant accomplishments should be included on the resumé to provide the greatest opportunity to be called for an interview.
6. List characteristics that are important to employers, including: leadership, organizational ability, good communication skills, problem solving, hard-working, and reliability.
7. Resumé templates seldom fit exact needs for specific jobs and should not be used.
8. Listing references is not required. At the bottom, state *References available on request.*
9. Proofread the resumé. It is a good idea to have someone else read the resumé before sending it.

Appearance

A resumé should be attractive and easy to read. It should fit on one page. Those with doctorates are exceptions; their resumés may be longer. Don't mix fonts; make sure the font and the size of the type are easily readable. Times New Roman is a good choice. Use only two sizes of type: one for the name and major headings and a smaller one for the rest. The margins should be one inch. Cut the number of words rather than shrinking the margins to squeeze in more words. Never staple a resumé; stapling makes it difficult to photocopy.

Example 1

Lloyd Bridges

Home address	2300 River Street,
	Hudson, NY 01234
	(231) 456-8910
School address	146 Oak Street,
	Rennselaer, NY 15678
	(423) 123-5678
email	bridge@RPI.edu

Experience:

Civil Construction Co., Syracuse,

NY (May–August 2007)
Leveling assistant in charge of ensuring level runways. Developed a simplified system of leveling.

Civil Engineering Dept., RPI, Rennsalaer,
NY (September 2006–May 2007)
Teaching assistant in CE 201, Strength of
Materials

Education:

Rennselaer Polytechnic Institute
B.S. Civil Engineering, expected May 2008
GPA 3.73
Relevant courses: Advanced Structural Analysis,
Concrete, Highway Construction
Project: Simulation of earthquake damage to
high-rise buildings

Skills:

CAD/CAM and other computer programs,
Weibull analyses, speaking knowledge
of Spanish

Honors:

3rd in class of 35. Dean's List (6 of 8 semesters)
Vice President, student ASCE chapter
(2007–2008)
Scholarship (2005–2008)

Activities:

Varsity ski team
Intramural football
Choir St Luke's Church in Albany
Served as Big Brother to young teen, 2007

Example 2

Cam Steel
202 Ford Street
Dearborn, MI 47172
(313) 145-6789
email: Steel@wsu.edu

Education
Wayne State University, B.S. Mechanical
    Engineering, expected 2009
GPA 3.36
Relevant courses: Strength of Materials, Physics,
    Computer Programming
Holy Cross High School, Dearborn, MI, 2005

Objective
Summer internship in automobile industry

Experience
Wayne State University Cafeteria (2006–2007),
    waiter
Joe's Auto Repair (2005–2007, after school),
    mechanic
High School debating team

Strengths
Hard-working, intelligent, organized, team player.
Willing to undertake any job.

Skills

Familiarity with UNIX, C, C++ computer
programs

Chapter Summary

Resumés should be short and direct. They should
emphasize the strengths of the candidate.

This concludes the formal presentation of
Reporting Results. The appendices that follow contain useful information on common errors in writing,
punctuation, and word choices to help develop writing skills. The appendices on the international system
of prefixes and units, on the Greek alphabet and its
typical uses, and on straight-line plotting of mathematical functions are provided as useful references in
technical writing.

Appendix I

Common Errors in Writing

This appendix is aimed at avoiding errors that the authors have seen in reading student papers and reviewing manuscripts for publication. Some of the examples are repeated from Chapter 1 for the convenience of the reader.

Pomposity

Avoid using large words where shorter ones would work just as well. For example, use *freezing* instead of *solidification*, *test* instead of *experimental investigation*, and *needs* instead of *requirements*.

Excessive Verbiage

A redundant word is an unnecessary word. Considering the high price of newsprint and book stock, we ought to watch for redundancies and pluck them from our writing as if we were picking ticks from a dog's back. Redundancies, like ticks, suck blood from our prose. Kilpatrick (1984)

Examples of excess verbiage are:

Example	*Instead use*
in order to	to
data points	data
at this point of time	now

Most uses of *respectively* can be eliminated without causing confusion. *Process* should be omitted in *casting process*, *machining process*, *rolling process*, etc.

Then can almost always be eliminated and should never be used more than once in the same paragraph. The following is a typical bad example: *The specimen was cut and then mounted in Bakelite. Then it was ground and polished.* Note that *The specimen was cut, mounted in Bakelite, ground, and polished* conveys the same meaning, but without the word *then*.

Avoid unnecessary redundancy. Instead of *try out, finish off, finally complete, absolutely necessary, triangular in shape*, and *very unique*, use *try, finish, complete, necessary, triangular*, and *unique*.

Appendix II

Punctuation

There is an excellent (and humorous) book, *Eats, Shoots and Leaves* by Lynn Truss (2003) that gives British rules for punctuation. The title emphasizes the importance of punctuation. The meaning of the title would be completely changed if the comma after *Eats* were omitted. Some excellent resources for American punctuation rules are *Webster's Standard American Style Manual* (1985) and *The ACS Style Guide: A Manual for Authors and Editors* (Dodd 1986). The following is a brief summary of punctuation that should serve as a starting point.

Commas

Commas have several uses. One is to substitute for *and* or *or* in a list of nouns. For example, *The experiment required a volt meter, some wire, an oscilloscope, and a battery* or *The alloy usually contains aluminum, titanium, or niobium.* Commas are also used with lists of adjectives. For example, *The flag was red, white, and blue.* The comma before the final *and* or *or* is called the *serial comma.* Americans tend to use the serial comma, whereas the British do not. The writer should pick one usage and be consistent.

Commas are also used to join separate clauses. *The volt meter was read after one minute, and the reading was recorded.*

Pairs of commas are used around a parenthetical comment such as *The engineer, a graduate of Cal Tech, was very clever* or *The author, L. H. Van Vlack, wrote many other texts.*

Commas are used between two equivalent modifying adjectives like *a strong, tough alloy* but not between two where the adjectives are very different, as in *an expensive steel beam.*

A comma should come before direct quotes, as in *Professor Allen asked, "Can anyone integrate this equation?"*

The placing or absence of commas can alter the meaning of a sentence. Compare *The steels, which were heat treated, proved satisfactory* with *The steels which were heat treated proved satisfactory.* The first implies that all the steels were heat treated, while the second states that only heat-treated steels were satisfactory. Or compare *The student claimed the professor was unfair* with *The student, claimed the professor, was unfair.*

Above all, the comma should be used to prevent ambiguity.

Hyphens

Hyphens also have several uses. One is to connect two nouns as in *aluminum-copper alloy* and *stress-strain curve.* Hyphens may also replace *versus* as in *stress-strain curves* or *volt-amperage relationship.*

Another use of hyphens is to connect two modifiers when one modifies the other and they act together as a single modifier for a noun. One example is *high-strength low-alloy steel*; here *high* modifies *strength* and *low* modifies *alloy.* Neither *high* nor *low* modifies *steel.* Another example is *plane-strain compression*, where *plane* modifies *strain* and together they modify *compression.* Note, however, that no hyphen is used in *compressed in plane strain.*

A third use of hyphens is to make words easier to pronounce. For example, *de-ice* is easier to read than *deice*, and *shell-like* easier than *shelllike.*

Still another use is to clarify meaning. For example, *re-mark* doesn't mean *remark* (or comment on); rather it means to mark again. Likewise, *re-formed* means formed again rather than *reformed*, which means turned from bad to good.

Hyphens are also used to break words that don't fit on a single line. For example, one might hyphenate: *In that respect it was a pains- taking job.* Note that care must be taken to make the break between pronounceable syllables. An incorrect example is: *In that respect it was a pain- staking job.* Other examples of correct breaks are *remem- ber* not *rememb- er* and *prin- ciples* not *pri- nciples.*

Often, numbers like *twenty-two* are hyphenated, but hyphenation is optional with words like *foot-pedal.*

Apostrophes

A noun with *'s* denotes possession. For example *the beam's strength* means *the strength of the beam* or *the machine's capacity* means *the capacity of the machine.* With plural nouns that end with *s*, the apostrophe follows the *s* as in *the beams' strengths,* meaning *the strengths of many beams,* or *the automobiles' mileage,* meaning *the mileage of many automobiles.* One exception to this is with proper names for which one pronounces the final *s*. The possessive of *Jones* is *Jones's* and of *Keats* is *Keats's.* However, a final *s* is not added with ancient proper nouns or proper nouns that are pronounced with a final *iz*, as in *Achilles' heel* and *Archimedes' screw.*

Apostrophes are also used in contractions to substitute for missing letters. For example, *The machine won't work* or *It's better to give than to receive.* Be careful to avoid the common mistake of confusing the possessive *its* with the contraction *it's*, meaning *it is.* Other common contractions are *can't, there's, haven't, we've,* and *he'd.* Although such contractions are acceptable in common speech and fiction writing, they should not be used in technical writing.

Quotation Marks

In American English, paired quotation marks (" ") are used for direct quotes, but not for paraphrasing of quotes. They can also be used in references around the title of an article or chapter. In addition, they can be used in text around a word or phrase that has questionable validity, as in *The professor said that an apostrophe was like a "bomb."*

In American English. for quotes within quotes, the inner quote is enclosed with single quote marks, as in *Mary said, "The professor quoted the book as saying 'Never assume that all equations are correct.'"* Note that in British English, the use of single and paired quote marks is reversed.

In American English, commas and periods always go inside quotation marks; the British put them outside unless they are part of the quote. Colons and semicolons always go outside quotation marks.

Colons and Semicolons

The rules for colons and semicolons are complicated. A colon is used as a mark of introduction when the clause, phrase, word, or series that follows the colon is linked to the preceding element. The most common use of a colon in technical writing is to introduce a list. For example, *An engineering decision must rest on a number of factors: short-term profitability, marketability, and safety.* Colons are also used in references to link subtitles to titles.

A colon may also introduce a phrase that explains, illustrates, amplifies, or restates the preceding phrase. For example, *Toughness is paramount in material choice for a pressure vessel: pressure vessels require high toughness.*

Semicolons are most often used to join two or more clauses when the second clause begins with a conjunctive adverb such as *accordingly, also, consequently, however, therefore,* or *thus,* as in the sentence *Stainless steel does not rust; therefore, it is used in the food industry, and it does not affect flavor.* However, the semicolon should not be used if the second clause is not closely related to the first. For example, the use of a semicolon in *Stainless steel finds application in the food industry; it contains at least 12 percent chromium* does not help explain why stainless steel is used in the food industry.

Semicolons can also be used to separate items in lists that contain internal commas. For example, *Metal Forming: Mechanics and Metallurgy*, third edition; *Materials Science: An Intermediate Text*; and *Materials for Engineers*, an undergraduate textbook.

Periods

Of course, a period indicates the end of a sentence. Periods are also used in abbreviations, such as *St. Venant's principle* and *et al.* Note that there is no period after *et* because it is a full Latin word meaning *and* but that *al.* is an abbreviation for *alia,* meaning *others.*

Italic Type

Italics are used in references to indicate a book title or a journal. Italics are used in mathematical expressions for variables. Note that in sin $(x/2)$ the variable x is italicized, but the abbreviation of sine function (sin) and the number 2 are not.

Italics are also often used to introduce and explain a new word or phrase that may be unfamiliar to the reader. They may also be used in examples, as is done throughout this book. Italics may also be used for emphasis, but they should not be overused for this purpose.

Brackets

There are four types of brackets, namely: parentheses ( ), square brackets [ ], braces { }, and angle brackets < >.

Parentheses are used for explanatory words or comments, as in *Hill's first anisotropic yield criterion (1948) was of a quadratic form...*, *stretcher strains (also known as Lüders bands) are...*, and *work-hardening (strain-hardening)....*

In mathematical usage, angle brackets come outside of braces, which are outside of square brackets, which are outside of parentheses, that is, { < [( )] > } or $z = <\mathrm{erf} \{\sin [1/(1-x)]\}>^2$. The writer can sometimes simplify complicated expressions by breaking them into multiple equations. For example, the preceding equation written as $z = [\mathrm{erf} (y)]^2$, where $y = \sin [1/(1-x)]$, is easier to read.

Ellipsis Points

An ellipsis, three consecutive periods, ..., is used for trailing off, as in mathematical series like $x + x^2/2! + x^3/3! + ...$, or to indicate missing words from a quotation.

Other Punctuation

Exclamation points should be avoided in technical writing. Question marks need no explanation, but they are rarely used in technical writing. Asterisks are sometimes used to designate footnotes. Bullets may be useful in oral presentations, but in the opinion of the authors have no place in technical writing.

Mathematical symbols, like $+$, $-$, $\neq$, $>$, $\geq$, $\leq$, and $/$, are used only in equations. The symbols @ and & should be avoided entirely, except in email addresses and in company names.

Capitalization

The first letter of the first word of a sentence is capitalized. A common mistake is to overcapitalize; within sentences only proper names should be capitalized. For example, *Poisson's ratio* is correct, not *Poisson's Ratio*. Names of elements are not capitalized, although the first letter of the symbol for a chemical element is. For example, *Cu* and *copper* are both correct.

Do not capitalize words following proper nouns. *Instron testing machine* is correct, but *Instron Testing Machine* is not.

For titles (figure titles, report titles, headings, etc.), capitalize either only the first word or all of the words except prepositions, articles, and conjunctions. For example, both *Schematic showing the drying concept* and *Schematic Showing the Drying Concept* are acceptable, but *Schematic showing the Drying Concept* is not.

Appendix III

Common Word Errors

There are many words that are easily confused with each other or that are commonly misspelled. Here is a collection that the authors find useful.

affect	v.t., produce an effect or influence
effect	v.t., to cause or accomplish; n., result or outcome
contaminate	v., to make impure
contaminant	n., a substance that contaminates
corroborate	v.t., to strengthen or confirm
collaborate	v.i., to cooperate or work with
discrete	adj., separate, disconnected
discreet	adj., tactful
ensure	v.t., to make sure something will happen
insure	v., to get protection

gage	n., test location of a tensile bar
gauge	n., instrument for measuring
grey / gray	grey is the British spelling of gray
plane	n., flat geometric surface; adj., flat, as in plane strain
plain	adj., simple or ordinary, as in plain carbon steel
principal	adj., first in rank, as in principal investigator or principal stress
principle	n., general truth or law
sample / specimen	a sample is a statistical group of specimens
silicon	n., element 28
silicone	n., a polymer with an S–O backbone
silica	n., SiO_2
stress	n., force per area
strength	n., critical value of stress, such as yield strength (not yield stress)

Spelling

It is *i* before *e* except after *c* and when sounding like *ay* as in *neighbor* or *weigh*. Many exceptions are contained in the sentence *The weird foreigner seizes neither leisure nor sport at its height.* Other exceptions include *either, being, obeisance, sheik, stein, counterfeit,* and *seismic.*

According to Henry Minott of the United Press, the fifteen most commonly misspelled words are:

changeable	dietitian	discernible	diphtheria
embarrass	gauge	harass	indispensable
judgment	likable	naphtha	occurred
paraphernalia	permissible	uncontrollable	

Other commonly misspelled words in technical writing are:

austenitizing	boundary	existence	foundry
height	inoculation	logarithm	martensite
regardless (not irregardless)	specimen	spheroidite	

Plural of Words of Greek or Latin Origin

Singular	Plural	Singular	Plural
analysis	analyses	appendix	appendices
colloquium	colloquia	criterion	criteria
datum	data	equilibrium	equilibria
focus	foci	index	indices
locus	loci	maximum	maxima
medium	media	minimum	minima
octahedron	octahedra	phenomenon	phenomena
tetrahedron	tetrahedra	thesis	theses
vacuum	vacua	vita	vitae

Use of Articles *a* and *an*

Whether *a* or *an* is used depends on the beginning sound of the following word or abbreviation. The article *a* is used before a consonant sound even if the word or abbreviation starts with a vowel. Examples are *a eutectic, a union, a U.S. senator, a one-time expense*, and *a UM professor*.

The article *an* is used before a noun or abbreviation that begins with a vowel sound even if the following word or abbreviation begins with a consonant. Examples are *an fcc lattice, an hour, an Rh factor, an n-p junction*, an MIT degree, *an unknown, an nth factorial*, and *an honor*.

Either *an* or *a* may be used before words that begin with a lightly stressed *h*. For example, *a history* and *an history* as well as *a heroic* and *an heroic* are acceptable.

Appendix IV

International System of Prefixes and Units

Table A.1. *Standard prefixes*

$10^3 n$	Name	Symbol
10^{-18}	atto	a
10^{-15}	femto	f
10^{-12}	pico	p
10^{-9}	nano	n
10^{-6}	micro	μ
10^{-3}	milli	m
10^3	kilo	k
10^6	mega	M
10^9	giga	G
10^{12}	tera	T

Table A.2. *Standard international system of units (SI)*

Symbol	Name	Quantity	Formula
A	ampere	electric current	base unit
Bq	becquerel	activity of a radio nuclide	1/s
C	coulomb	electric charge	A•s
°C	degree Celsius	temperature interval	base unit
cd	candela	luminous intensity	base unit
F	farad	electric capacitance	C/V
Gy	gray	absorbed dose	J/kg
g	gram	mass	kg/1000
H	Henry	inductance	Wb/A
Hz	Hertz	frequency	1/s
ha	hectare	area	$10{,}000 \text{ m}^2$
J	joule	energy, work, heat	N•m
K	Kelvin	temperature	base unit
kg	kilogram	mass	base unit
L	liter	volume	$m^3/1000$
lm	lumen	luminous flux	cd•sr
lx	lux	illuminance	lm/m^2
m	meter	length	base unit
mol	mole	amount of substance	base unit
N	Newton	force	$kg•m/s^2$
Pa	Pascal	pressure, stress	N/m^2
rad	radian	plane angle	(dimensionless)
S	Siemens	electric conductance	A/V
s	second	time	base unit
sr	steradian	solid angle	(dimensionless)
Sv	sievert	dose equivalent	J/kg
T	Tesla	magnetic flux density	Wb/m^2
t	tonne, metric ton	mass	1000 kg; Mg
V	volt	electric potential	W/A
W	ohm	electric resistance	V/A
W	watt	power, radiant flux	J/s
Wb	Weber	magnetic flux	V•s

Appendix V

The Greek Alphabet and Typical Uses

Greek letters are frequently used for technical variables. The following table shows the most common usages for the Greek letters.

Letter	Symbol	Typical Use
alpha	A	
	α	angle, coefficient of thermal expansion
beta	B	
	β	angle
gamma	Γ	mathematical function
	γ	angle, shear strain, surface energy
delta	Δ	difference
	δ, ∂	difference between differential quantities
epsilon	E	
	ε	strain
zeta	Z	
	ζ	
eta	H	
	η	viscosity, efficiency

(continued)

Letter	Symbol	Typical Use
theta	Θ	temperature
	θ	angle
iota	I	
	ι	
kappa	K	
	κ	
lambda	Λ	
	λ	wavelength
mu	M	
	μ	coefficient of friction, 10^{-6}
nu	N	
	ν	frequency, Poisson's ratio
xi	Ξ	
	ξ	
omicron	O	
	o	
pi	Π	multiplicative series
	π	3.1415926…
rho	P	
	ρ	density, radius of curvature
sigma	Σ	summation
	σ	stress, conductivity, standard deviation
tau	T	
	τ	shear stress
upsilon	Υ	
	υ	
phi	Φ	
	ϕ	angle
chi	X	
	χ	
psi	Ψ	
	ψ	angle
omega	Ω	ohm, the end
	ω	angular frequency

Appendix VI

Straight-Line Plots for Some Mathematical Functions

Plotting experimental data on scales chosen so that the theory gives a straight line allows one to find constants in mathematical expressions. For example, if data are to be fitted to $y = ax + b$ in a plot (see Figure A.1) of y vs. x, b is the value of y at $x = 0$ and a is the slope.

Figure A.1. Data (x_i, y_i) plotted on linear scales may be fitted to the equation $y = ax + b$, where b is the value of y at $x = 0$ and a is the slope.

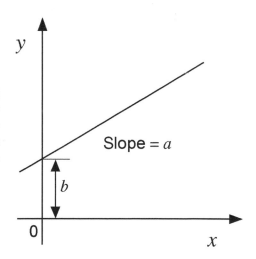

Slope $= a$

The following examples are adapted from the fifth edition (1989) of J. P. Holman's book *Experimental Methods for Engineers*. All of the plotting methods produce a straight line on linear scales and thus facilitate least-squares fitting routines to determine a best-fit straight line. In some of the examples the first paired points (x_1, y_1) are used in plotting the ordinate values.

Function	Plot	Graph
$y = ax^b$	$\ln y$ vs. $\ln x$	
$y = a\exp(bx)$	$\ln y$ vs. x	
$y = 1 - \exp-(bx)^n$	$\ln\ln\left(\dfrac{1}{1-y}\right)$ vs. $\ln x$	

Function	Plot	Graph
$y = a\exp(bx + cx^2)$	$\dfrac{\ln\,(y/y_1)}{x - x_1}$ vs. x	
$y = a + bx + cx^2$	$\dfrac{y - y_1}{x - x_1}$ vs. x	
$y = \dfrac{a}{x} + b$	y vs. $\dfrac{1}{x}$	
$y = \dfrac{x}{a + bx}$	$\dfrac{1}{y}$ vs. $\dfrac{1}{x}$	

Function	Plot	Graph
$y = \dfrac{x}{a + bx} + c$	$\dfrac{y - y_1}{x - x_1}$ vs. x	
$y = b + a\sqrt{x}$	y vs. $\sqrt{x}$ or y^2 vs. x	

References

M. F. Ashby and D. R. H. Jones (1981), *Engineering Materials 1: An Introduction to Their Properties and Applications*, Butterworth/Heinemann, Oxford.

W. Chauvenet (1863/1961), "A Manual of Spherical and Practical Astronomy," Vol. II, *Theory of Astronomical Instruments: Method of Least Squares*, J. B. Lippincott, Philadelphia/Peter Smith Publisher, New York, pp. 564–66.

G. E. Dieter (1991), *Engineering Design: A Materials and Processing Approach*, second edition, McGraw-Hill Publishing Co., New York.

J. S. Dodd, ed. (1986), *The ACS Style Guide: A Manual for Authors and Editors*, American Chemical Society, Washington, D.C.

J. F. Hatch (1984), *Aluminum: Properties and Physical Metallurgy*, ASM, Metals Park, Ohio.

J. P. Holman (1989), *Experimental Methods for Engineers*, fifth edition, McGraw-Hill Book Co., New York, p. 63.

L. G. Johnson (1951), "The Median Ranks of Sample Values in Their Population with an Application to

Certain Fatigue Studies," *Industrial Mathematics*, vol. 2., pp. 1–9.

J. J. Kilpatrick (1984), *The Writer's Art*, Andrews, McMeel and Parker, Kansas City, Mo.

S. J. Kline and F. A. McClintock (1953), "Describing Uncertainties in Single-Sample Experiments," *Mechanical Engineering*, January, p. 3.

P. B. Schubert (1953), *Pipe and Tube Bending*, Industrial Press, New York.

L. Truss (2003), *Eats, Shoots and Leaves*, Profile Books, London.

Webster's Standard American Style Manual (1985), Merriam-Webster Inc., Springfield, Mass.

Index